SpringerBriefs in Molecular Science

Chemistry of Foods

Series editor

Salvatore Parisi, Industrial Consultant, Palermo, Italy

The series Springer Briefs in Molecular Science: Chemistry of Foods presents compact topical volumes in the area of food chemistry. The series has a clear focus on the chemistry and chemical aspects of foods, topics such as the physics or biology of foods are not part of its scope. The Briefs volumes in the series aim at presenting chemical background information or an introduction and clear-cut overview on the chemistry related to specific topics in this area. Typical topics thus include: - Compound classes in foods – their chemistry and properties with respect to the foods (e.g. sugars, proteins, fats, minerals, …) - Contaminants and additives in foods – their chemistry and chemical transformations - Chemical analysis and monitoring of foods - Chemical transformations in foods, evolution and alterations of chemicals in foods, interactions between food and its packaging materials, chemical aspects of the food production processes - Chemistry and the food industry – from safety protocols to modern food production The treated subjects will particularly appeal to professionals and researchers concerned with food chemistry. Many volume topics address professionals and current problems in the food industry, but will also be interesting for readers generally concerned with the chemistry of foods. With the unique format and character of Springer Briefs (50 to 125 pages), the volumes are compact and easily digestible. Briefs allow authors to present their ideas and readers to absorb them with minimal time investment. Briefs will be published as part of Springer's eBook collection, with millions of users worldwide. In addition, Briefs will be available for individual print and electronic purchase. Briefs are characterized by fast, global electronic dissemination, standard publishing contracts, easy-to-use manuscript preparation and formatting guidelines, and expedited production schedules. Both solicited and unsolicited manuscripts focusing on food chemistry are considered for publication in this series.

More information about this series at http://www.springer.com/series/11853

Pasqualina Laganà · Emanuela Avventuroso
Giovanni Romano · Maria Eufemia Gioffré
Paolo Patanè · Salvatore Parisi
Umberto Moscato · Santi Delia

Chemistry and Hygiene of Food Additives

Springer

Pasqualina Laganà
Department of Biomedical and Dental Sciences
   and Morphofunctional Imaging
University of Messina
Messina
Italy

Emanuela Avventuroso
Department of Biomedical and Dental Sciences
   and Morphofunctional Imaging
University of Messina
Messina
Italy

Giovanni Romano
Territorial Dietician
Lamezia Terme
Italy

Maria Eufemia Gioffré
Food Safety Consultant
Messina
Italy

Paolo Patanè
University of Messina
Messina
Italy

Salvatore Parisi
Industrial Consultant
Palermo
Italy

Umberto Moscato
Department of Public Health
Università Cattolica del Sacro Cuore
Rome
Italy

Santi Delia
Department of Biomedical and Dental Sciences
   and Morphofunctional Imaging
University of Messina
Messina
Italy

ISSN 2191-5407   ISSN 2191-5415   (electronic)
SpringerBriefs in Molecular Science
ISSN 2199-689X   ISSN 2199-7209   (electronic)
Chemistry of Foods
ISBN 978-3-319-57041-9   ISBN 978-3-319-57042-6   (eBook)
DOI 10.1007/978-3-319-57042-6

Library of Congress Control Number: 2017937704

Printed on acid-free paper

This Springer imprint is published by Springer Nature
The registered company is Springer International Publishing AG
The registered company address is: Gewerbestrasse 11, 6330 Cham, Switzerland

Contents

3 Food Additives and Effects on the Microbial Ecology
Pasqualina Laganà, Emanuela Avventuroso, Giovanni Romano,
Maria Eufemia Gioffré, Paolo Patanè, Salvatore Parisi,
Umberto Moscato and Santi Delia

4 Use and Overuse of Food Additives in Edible Products:
Pasqualina Laganà, Emanuela Avventuroso, Giovanni Romano,
Maria Eufemia Gioffré, Paolo Patanè, Salvatore Parisi,
Umberto Moscato and Santi Delia

About the Authors

Pasqualina Laganà is a Biologist with a M.Sc. from the University of Messina, Italy, and specialisations in General Pathology (1992) and Chemistry and Food Technologies (2000). Graduated in 'Risks and disorders in the Work' (1998) and in 'Parasitology of Territory' (2009), she has also obtained in 2013 a Master in 'Epidemiology and Biostatistics' (Università Cattolica del Sacro Cuore, Rome, Italy). Since 1996 she has been Adjunct Professor of General and Applied Hygiene, and Researcher at the Faculty of Medicine and Surgery at the University of Messina. Member of the Italian Society of Hygiene, Preventive Medicine and Public Health (S.It.I.), and the Italian Study Group Hospital Hygiene and Italian Foods Group, she is also the Responsible of Analysis within the Regional Reference Laboratory for clinical and environmental Legionella Surveillance (branch of Messina, Sicily) since 2012, and the Head of the Laboratory from October 2016 onwards. She is a member of the Editorial Board of several Journals, including the Journal of Clinical Microbiology and Biochemical Technology (JCMBT), Zeszyty Naukowe (Gdynia Maritime University, Poland), Global Drugs and Therapeutics (GDT) and EC Microbiology (ECMI). She is also a Referee for various international scientific journals and the author of 70 publications indexed in Medline and other databases.

Emanuela Avventuroso is a Medical Surgeon graduated from University of Messina, Italy. Since 2012 she is a Doctor in Specialist Training in Hygiene and Preventive Medicine at the University of Messina. Dr. Avventuroso had the opportunity to deepen some of the different areas within Hygiene and Public Health, particularly in the early years of her specialisation in Food and Environmental Hygiene, and then taking care of Vaccinations, Infectious Diseases and Health Management.

Giovanni Romano is a dietician with a M.Sc. cum laude at the University of Messina, Italy, and he holds a 'sport nutritionist expert' certification from the SANIS School, Italian National Olympic Committee (CONI), Palermo, Italy. He also holds an international certification for sport nutrition and supplementation. Dr. Romano is currently responsible for the 'Nutrition and Diet Therapy' in a Private Bio-Medical Office, Lamezia Terme, Italy. He is also member of the 'International Society of Sport Nutrition' (USA) and the nutrition expert of provincial AVIS, Catanzaro, Italy. In addition, Dr. Romano has lectured sport nutrition for the Italian pilot project 'Muoviamoci Insieme' in cooperation with CONI Calabria, Italy.

Maria Eufemia Gioffré graduated in Biology Science at the Food Safety Consultant, Italy, in 2008. She has qualified from the same university as Biologist. She obtained the Ph.D. in 'Applied Hygiene' in April 2012 from the same university, and currently, she works as a Quality Control Manager for a food company, with other professional consultancies in the food sector. Dr. Gioffré has given important scientific contributions in the fields of Food Hygiene and Food Microbiology.

Paolo Patanè is graduated in law, with top marks, from the University of Messina, Italy, and he has obtained an M.Sc. in Innovation and Development of Intellectual Property in Rome. After the successful Italian state examination to become a lawyer, Dr. Patanè has worked as consultant for legal and administrative aspects of the Sicilian Regional Authority, being in charge of Management EU funds in the Operating Regional Programme 'POR Sicilia 2000–2006'. He was also General Secretary of the Faculty of Medicine and Surgery at the University of Messina (2007–2013). After a short period at the Department of Clinical and Experimental Medicine (2013–2015), he is now the Head of the Office of the University of Messina, and responsible for protecting the Intellectual Property Rights of the inventions from the university staff. He has also been a University contract professor (2013/2014 and 2014/2015) at the Messina University, and a lecturer in 2015 and 2016 at seven workshops about Intellectual Property Rights. Dr. Patanè has also published several issues about law ("Authority", "Ombudsman", "Spoils System", "Prisoners population in Italy").

Salvatore Parisi is a chemist and food scientist, working as a consultant for the food industry and as a lecturer in the academia. He obtained his M.Sc. from the University of Palermo, and Ph.D. from the University of Messina, Italy. Dr. Parisi is also a Preventive Controls Qualified Individual (PCQI) after the successful completion of the FSPCA Preventive Controls for Human Food course in 2016 (according to new U.S. regulations). He serves as series editor for the SpringerBriefs in Molecular Science: Chemistry of Foods and he is a member of the Editorial Board of different scientific journals. Dr. Parisi is an associate of the Sicilian Order of Research Chemists, Italy, and many other international Associations, including the EFSA's Expert Database, Parma, Italy, and the FAO Food Safety Expert Roster, FAO, Rome, Italy. In addition, Dr. Parisi is a member of the AOAC Official Methods of Analysis SM (OMA) Expert Review Panel (ERP) for Fertilizers. He has important scientific contribution (with more than 120 articles, books and software) in the fields of Packaging Technology, Polymer Chemistry, Shelf Life Prediction and Food Microbiology (Italian Oxoid Award, 2001).

Prof. Umberto Moscato is a Medical Doctor, specialist in Hygiene, and Associate Professor of General and Applied Hygiene at Università Cattolica del Sacro Cuore, Rome. He currently holds several positions as a professional consultant for public and private healthcare facilities for welfare services. He is a member of several scientific societies and, in particular, a member (and now President) of the Scientific Board of the Italian Society of Hygiene, Preventive Medicine and Public Health (S.It.I.). He is a Referee and Member of various national and international organisations and scientific journals, including the Italian Ministry of Health, the World Health Organization, the European Center of Control Disease, Annals of Hygiene, Prevention Today, and BioMed Central Journals. Since 2003 he is a member of the Editorial and Scientific Board of the Italian Journal of Public Health. At present, Prof. Moscato has different assignments by Italian and International Institutions, including the Faculty of Medicine and Surgery 'A. Gemelli' and the Polytechnic University of Milan. His research activity covers new analytical methods in biochemistry, physics, and toxicology as is documented in approximately 252 publications indexed in Medline and other databases, as well as congress extended papers, journal articles, congress abstracts, monographs, books and book chapters.

Prof. Santi Delia is Full Professor in the Department of Hygiene at University of Messina, Italy. After an initial scholarship in 1974, he was appointed as an assistant in the same department since 1976, and conducted other tasks related with hygienic aspects. During this period he was the Coordinator of the CIO Operative Group and Chairman of the Commission Surveillance for Food Arrangements, and responsible for two official teachings in the Degree Course of Medicine and Surgery. Also, he has lectured in different postgraduate schools for various degree programs. Professor Delia was the Director of the Master course in 'Food Hygiene and Food Legislation' and the Coordinator of three cycles of PhD School in 'Applied Hygiene'. He was also the Director of the Department of Hygiene, Preventive Medicine and Public Health 'R. De Blasi' (University of Messina, 2009–2012) and the Director of the Food Laboratory and the Regional Reference Laboratory for Environmental and Clinical Surveillance of Legionellosis until September 2016. Professor Delia has taken part in numerous conventions, conferences, roundtables, advanced training courses, as a speaker and/or moderator. His scientific activity is documented in more than 150 scientific articles, books and book chapters.

Chapter 1
Classification and Technological Purposes of Food Additives: The European Point of View

Pasqualina Laganà, Emanuela Avventuroso, Giovanni Romano, Maria Eufemia Gioffré, Paolo Patanè, Salvatore Parisi, Umberto Moscato and Santi Delia

Abstract The aim of this chapter was to describe the regulatory classification and the list of 'technological purposes' of food additives by the European viewpoint. At present, food additives are extremely useful. Significant progress has been made in the field of health and nutrition. Food additives play an important role in the complex chain of modern food productions. In general, foods are subject to many environmental variables which may alter the original composition. Moreover, the presence of additives in foods must be declared on the label as an ingredient. Also, all additives used in the food industry are included in a 'positive list' by health organisations. Finally, food additives are classified into functional categories, in relation to the declared action: food protection, colour or flavour enhancement, sugar substitution or structural and technological functions. This chapter provides also a description of these sub-categories with correlated examples.

Keywords Acceptable daily intake · Dye · European Food Safety Authority · Food additive · Preservative · Sweetener · Thickener

Abbreviations

ADI	Acceptable daily intake
AFC	Panel on food additives, flavourings, processing aids and materials in contact with food
ANS	Panel on Food Additives and Nutrient Sources Added to Food
BHA	Butylated hydroxyanisole
BHT	Butylated hydroxytoluene
CEF	Panel on Food Contact Materials, Enzymes, Flavourings and Processing Aids
EDTA	Ethylenediaminetetraacetic acid
EFSA	European Food Safety Authority
EU	European Union
IMF	Intermediate moisture
INRAN	Italian National Research Institute for Food and Nutrition
INS	International Numbering System

© The Author(s) 2017

P. Laganà et al., *Chemistry and Hygiene of Food Additives*, Chemistry of Foods, DOI 10.1007/978-3-319-57042-6_1

JECFA   Joint FAO/WHO Expert Committee on Food Additives
NOAEL   No Observed Adverse Effect Level
O$_2$ Oxygen
PHB   *p*-hydroxybenzoic acid
SO$_2$ Sulphur dioxide
TBHQ   Tertiary butyl hydroquinone
USA   United States of America
WHO   World Health Organization

1.1   Food Additives. An Introduction

Food additives are still one of the most controversial topics in the modern world and may represent a concern for the consumer. However, food additives have been used for centuries, being associated with the most recent food production technologies. The birth of food preservation techniques may be associated with ancient crop storage and the use of salting and smoking methods for meat and fish. The Jews used the water of the Dead Sea in 1600 BC; the Egyptians and the Romans were aware of the disinfectant power of sulphur dioxide when speaking of winemaking productions. These techniques, used today with the aid of advanced technology, were based on the correlation between the water content in foods and correlated durabilities. It is known that the Egyptians and Romans used dyes, flavourings, spices and salt to make more tasty food and to prolong storage life (Saltmarsh 2000). The preparation and food storage are objectives shared by the traditional cuisine and the industry. In the last 50 years, the developments of food science and technology and the concomitant evolution of consumers' requests led to a substantial increase in the use of food additives, combined with new industrial technologies.

At present, food additives are extremely useful as powdered mixtures for sauces, prepared dishes for instant ready meals and snacks. Significant progress has been made in the field of health and nutrition: in fact, many food products available today could not exist without food additives.

1.2   Basic Definitions and Requirements for Food Additives

Given the great importance of food additives, it is necessary to define these substances on the basis of the declared justification for use (sensory, technological or nutritional purposes). In the European Union (EU), food additives are defined as specified by the Council Directive 89/107/EEC of 21 December 1988 on the approximation of the laws of the Member States concerning food additives authorised for use in foodstuffs intended for human consumption. The exact definition is: '... *Any substance not normally consumed as a food in itself and not normally used as a characteristic ingredient of food, regardless of whether it has nutritive value, the intentional addition of which to food for a technological purpose in the manufacture, processing, preparation, treatment, packaging, transport or storage of foods, it may be reasonably expected to become, in it or its derivative, a component of such foods, either directly or indirectly*' (Council of the European Communities 1988). In addition, it should be noted that:

(a) Additives for food production must be safe
(b) The use of these chemicals or substances should not cause any damage, or present a risk of toxicity to humans or animals intended for human consumption
(c) Food additives do not react with any of the constituents of the food; also,
(d) Compounds associated with possible teratogenic, mutagenic and carcinogenic risks are strictly prohibited
(e) All food additives must have a high degree of purity, according to the standards of the current legislation (Mariani and Testa 2009).

Food additives play an important role in the complex chain of modern food productions. In general, foods are subject to many environmental variables, such as temperature fluctuations, oxidation and exposure to microbes, which may alter the **original** composition (Ministero della Salute 2012). The use of these substances is essential to preserve food quality and features. Moreover, the presence of additives in foods must be declared on the label as an ingredient. Also, all additives used in the food industry are included in a 'positive list' by health organisations that provide for timely and continuously update (European Parliament and Council 2010).

1.3 Classification of Food Additives

At present, the use of food additives should be conducted according to well-defined 'good employment practices' rules. The required mention of these substances on food labels is correlated with their classification in accordance with the International Numbering System (INS): a peculiar identification code is assigned to all EU approved additives, consisting of the letter 'E' followed by 3 or 4 digits. However, each additive may be also indicated on the label with their chemical name.

Many additives have different functions, belonging also to multiple or 'functional' categories; for these reasons, the main use has to be declared. Moreover, a negative list containing a list of substances with proven toxic or harmful effects on human health has been created: the use of chemicals mentioned in this list is expressly prohibited.

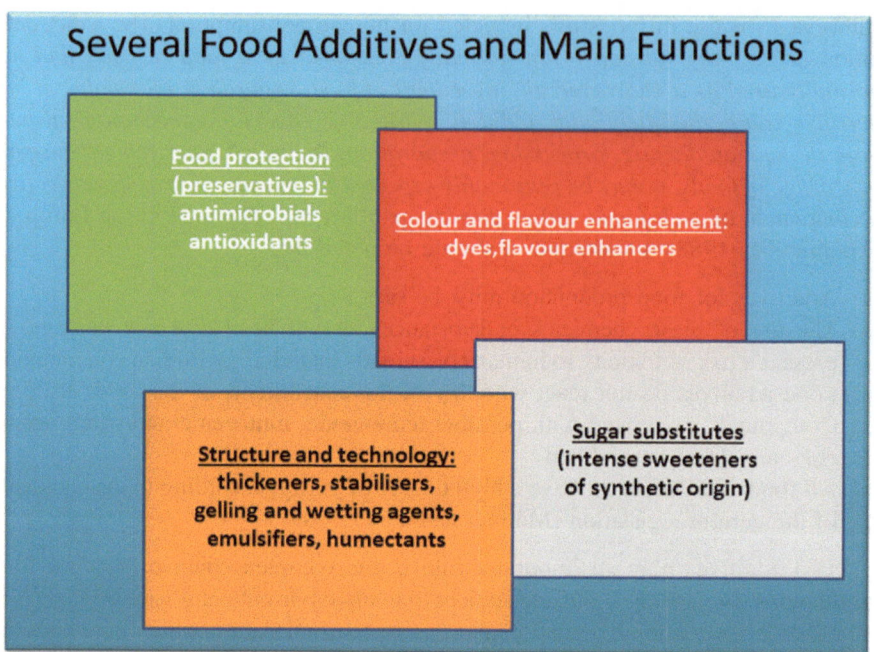

Fig. 1.1 A simplified subdivision of several food additives with relation to main functions. Certain additives (coating agents, propellants, melting salts, gases for modified atmosphere treatments) are not shown here

As above mentioned, food additives are classified into functional categories, in relation to the declared technological function as defined in the Regulation (EC) No 1333/2008 (Saltmarsh 2000). Basically, functions of food additives can be summarised in four major areas:

- Food protection: antimicrobials and antioxidants
- Dyes: recovery or enhancement of natural colours; achievement of special colours
- Sugar substitution (intense sweeteners)
- Structure and technology: thickeners, gelling agents, stabilisers.

A simplified classification of possible sub-categories is shown in Fig. 1.1. Certain additives (coating agents, propellants, melting salts, gases for modified atmosphere treatments) are not shown here. The following sections provide a description of these sub-categories with a single example wherever possible.

Table 1.1 Some examples concerning preservatives and possible used in foodstuffs

Code	Substance/category	Foods
E 200–203	Sorbic acid and sorbates (sodium sorbate, E 201, is not allowed in the European Union)	Cheese, wine, nuts, fruit-based sauces
E 210–213	Benzoic acid and benzoates	Sauces, pickled vegetables, jams and low-sugar jellies, candied fruit, fish-based products
E 220–228	Sulphur dioxide and sulphites	Wine, nuts, potatoes
E 235	Natamycin	Surface treatment of vegetables and sausages
E 249–252	Nitrites and nitrates	Meat, ham, bacon, pickled herrings, cheese

1.3.1 Food Preservatives

Chemicals with antimicrobial features can be found easily when speaking of acids, bases, biphenyls, heavy metals and halogens. Many of these compounds are toxic and therefore cannot be used in the food industry. On the other hand, there are substances with positive effects on the shelf life of foods.

The ability to avoid microbially caused alterations may have important economic consequences. Some foods are an excellent substrate for the microbial development: pH, temperature and water activity are important elements. For these reasons, the use of chemical preservatives is required when physical treatments such as freezing, sterilisation and dehydration cannot be carried out.[1] These compounds are usually divided into two categories: natural and synthetic preservatives.

Natural food preservatives such as acetic acid or sodium chloride are usual components of foods; they may be also obtained as a result of fermentation processes. With relation to fermentation products, lactic and acetic acids are well known: these compounds can act as antiseptic substances, lowering pH, and modify flavour and aroma of the finished product. Other natural compounds added to foods for several purposes are oils, vinegar, ethyl alcohol and sodium chloride; there is no specific legislation for these natural preservatives.

In contrast, artificial chemical preservatives—listed from E 200 to E 297—perform different functions. Their use is allowed on condition that a specific legislation is taken into account, particularly with reference to permitted concentrations in foods (European Parliament and Council 2010; Mariani and Testa 2009). Some examples of preservatives and known uses are shown in Table 1.1.

---

[1]Marina et al.

1.3.1.1 Synthetic Preservatives

Synthetic preservatives are divided into three major groups depending on their health effects and the declared use. Some data about the potential toxicity of antimicrobial compounds for food production purposes are listed in Table 1.2

'Harmless' Preservatives

These substances are considered harmless because they are metabolised without the production of toxic derivatives. Examples: propionic acid, sorbic acid and its salts (Mariani and Testa 2009). An interesting compound, proposed as a broad-spectrum preservative, is *N*-alpha-lauroyl-L-arginine ethyl ester monohydrochloride (European Commission 2015; Hawkins et al. 2009; Ruckman et al. 2004); uses: fruit nectars, fruit, vegetables, baked goods, meat products, fish, sauces.

Preservatives Used in Relation to Risks and Advantages

Certain preservatives are used after a careful evaluation of risks and advantages, although specific prohibitions may be judged appropriate depending on the direct or indirect toxic action.

The following chemicals can be defined as good examples when speaking of risk/benefit evaluation:

(1) Formaldehyde. This preservative, used as a bacteriostatic agent against butyric fermentation in *Grana* cheese, may be not used because butyric 'swelling' can be prevented with accurate milk hygiene. In addition, the addition of lysozyme may be useful. Formaldehyde has toxic other harmful effects; moreover, it is also a neurotoxin (EFSA ANS 2014). At present, the EU considers a maximum limit of 50 mg/kg in foods provided that residues do not affect consumers'

Table 1.2 Antimicrobial agents for food production and related toxicity

Additive code	Chemical name	Possible toxicity
E 210	Benzoic acid	Toxic substance with low ADI (5 mg/kg body weight)
E 214	Ethyl-*p*-hydroxybenzoate	As above mentioned
E 220	Sulphur dioxide	This quite toxic substance interacts with metabolism and seems to inactivate vitamin B$_1$
E 230	Diphenyl	A very toxic substance with low ADI (0.005 mg/kg body weight)
E 235	Piramicyn	Toxic substance
E 236	Formic acid	Toxic substance

health. The use of formaldehyde as hexamethylenetetramine (E 239) is currently permitted with a maximum residue level of 25 mg/kg in *Provolone* cheese, according to Directive No 95/2/EC

(2) Ethylene oxide is used as a sterilising agent for spices and vegetable drugs. This compound reacts with water in foods with the consequent production of ethylene glycol (1,2-ethanediol): this compound can turn into chloroethanol (ethylene chlorohydrin) in the presence of sodium chloride. Alternatively, propylene oxide can be used. Ethylene glycol has a depressant action on the nervous system, causing seizures and kidney damage

(3) Sulphur dioxide (SO_2) has to be considered in this ambit when speaking of risk/benefit ratio. This preservative—coded E 220—is essential in winemaking (no alternative products are available). SO_2 is found in two forms: free and combined sulphur dioxide. The free form inhibits the microbial spreading slowing down fermentation processes (in particular, the development of *Kloeckera epiculata* is inhibited. SO_2 is substantially an anti-thiamine agent (it breaks the two-ring system of vitamin B_1). Consequently, the nutritional value of foods containing vitamins B_1 and B_{12} is reduced

(4) Sulphites (sodium and potassium salts) are coded as E 221 and E 225 respectively—are not toxic and can be turned into inert compounds in the human body. However, they can react with disulfide bonds of proteins reducing them to SHY groups, with consequent structural changes in cellular enzymes and vitamins. Sulfites react with the intestinal microflora generating toxic products (EFSA CEF 2011; Garcia-Fuentes et al. 2015). Because of antioxidant properties, SO_2 and sulphites are used mainly in wine and beer, and in other products (juices, ciders, vinegar, mix flour and potato flakes). It has to be noted that sulphites are considered dangerous for people with asthma (maximum limit: 10 mg/kg and the European Commission believes that the use of SO_2 should be reduced to a minimum level (Mariani and Testa 2009).

(5) Certain substances may be used with the aim of increasing shelf life of foods; however, the use of these substances must be safe and the regular updating of additive lists is requested (EFSA AFC 2006). Certain toxic ingredients, widely used in the past, have been prohibited at present: salicylates and borates are good examples. In detail, salicylic acid causes stomach ulcers, while boric acid and borates cause nausea, headache, headaches, insomnia, alopecia and skin eruptions

(6) Benzoic acid and benzoates are excellent antimicrobial agents with maximum effectiveness for pH values comprised between 2.5 and 4.0. However, the World Health Organization (WHO) has questioned in 2000 the uniqueness of benzoates, and some Australian studies have correlated the use of these compounds in foods intended for children with the onset of the 'Attention Deficit Hyperactivity Disorder' (ADHD) syndrome. The European Food Safety Authority (EFSA) is carrying out research to explore the topic and its implications on human health (Mariani and Testa 2009)

(7) Esters of *p*-hydroxybenzoic acid (PHB) are considered acceptable food preservatives: their use as a food additive is permitted as a result of the evaluation of the Joint FAO/WHO Expert Committee on Food Additives (JECFA). The 'No Observed Adverse Effect Level' (NOAEL) for PHB additives is not defined at present: consequently, E 216 and E 217 additives have been removed by Commission Regulation (EU) No 1129/2011. Parabens are used as preservatives in cosmetics with limitations and restrictions that may vary from Country to Country.

Compounds for Surface Treatment

Some anti-mould agents are antibiotic substances: they may be used for mass or surface treatment. This group includes various substances such as diphenyl, thiabendazole and neomycin.

Natamycin can be used for surface treatments of cheeses, while nisin (E 234) is recommended for the treatment of cheese mass, canned vegetables and creams. Other anti-mould substances are carbon dioxide (E 290) for post-harvest treatments of fruits and vegetables and thiabendazole. The European Parliament has defined in 2005 maximum amounts for nicin and piramicyn in foods. Nisin is admitted in semolina and tapioca puddings up to 3 mg/kg, and in cheese up to 12.5 mg/kg (EFSA AFC 2006). Natamycin can be added in the amount of 1 mg/dm^2 and up to a maximum of 5 mm in depth when speaking of hard cheeses, hard seeds and seasoned sausages (Honikel 2008). Natamycin can be used for citrus fruits and correlated packaging materials (its action is mainly against *Penicillium* fungi). Notwithstanding all precautions, a small part of the preservative can also penetrate into the fruit, but this should not pose any problems to the consumer. The literature reports that workers employed in the field of shipments of citrus (by ship or truck), then exposed for long periods to this preservative, showed some sensitivity to diphenyl, complaining of allergic reactions, nausea, vomiting, eye irritation and to the nasal mucosa. There are no significant side effects because the diphenyl is a compound that is excreted by the kidneys in unchanged form. However, the use of this food additive is phased out in Australia because of possible sensitivity to diphenyl: observed effects are allergic reactions, nausea, vomiting, eye irritation and damages to the nasal mucosa (Mariani and Testa 2009).

1.3.1.2 Secondary Preservative Action—Nitrites and Nitrates

Nitrites and nitrates are obtained by nitrous and nitric acid respectively. The subtle difference between nitrite and nitrate anions is the presence of a single atom of oxygen in nitrates. Sodium nitrate is the most widely nitrogen-based compound used on earth (Honikel 2008). As food additives, nitrites are identified by the initials E 249 and E 250, while E 251 and E 252 are for nitrates. The use of nitrites and nitrates in foods exceed the function of simple preservatives (the dosage would be much lower than those used) because these substances may be also act as processing aids to keep the colour of meats. In fact, these substances are added to sausages, hams, tinned meat and other products based on meat, marinated fish and sometimes dairy products. Nitrites and nitrates are used for the following reasons:

(1) Red colours of meats should be preserved
(2) Certain aromas should be enhanced or promoted, with a selective action against microorganisms that determine the maturation of meats
(3) Antiseptic and antimicrobial effects are well known, especially against *C. botulinum*.

Nitrates have by law a maximum limit of 250 ppm in foods as a human target, while the limit is 150 ppm for nitrites, having regard to the greater toxicity. Nitrite shows selective action on microorganisms (Honikel 2008). It is capable of inhibiting various enzymatic systems and respiratory chains, especially by oxidising the iron atoms of many enzymes. The primary function of the nitrate as a preservative is related to the ability to inhibit *Clostridium botulinum*. However, the concentration of nitrate capable of blocking clostridia is lower than the real added substance, and is partly replaced by ascorbic and sorbic acids. The addition of nitrite up to 100 mg/kg may be sufficient for some foods, although concentrations up to 150 mg/kg may be required in certain applications. The EFSA agrees the total amount of added nitrates and nitrites is important when speaking of food safety in meat products and traditional salting (Mariani and Testa 2009).

The European law allows the addition in foodstuffs of a maximum quantity of 150 mg per kg of product for nitrites, 25 times the maximum amount in vegetables. The main danger of nitrites is their ability to react with amines (organic compounds abundant particularly in protein foods such as meat, cold cuts and cheeses), generating nitrosamines, and potentially carcinogenic substances. Consequently, the Scientific Committee for Food of the European Commission assessed the acceptable daily intake of nitrites and nitrates as 0.06 and 3.7 mg/kg of body weight respectively.

1.3.2 Antioxidants

Antioxidant compounds are substances able to prolong the shelf life of a food product because of their effects against oxidative reactions (consequences: fat rancidity, colorimetric variations, etc.). Antioxidants are active at very low concentrations. Some substances maintain their antioxidant power even after cooking treatments and their effects on fat matter. Antioxidants are coded starting with E 300 (L-ascorbic acid). Rancidity and food browning can alter both solid and liquid products with consequent unacceptability. These alterations may have chemical–physical origins (light, temperature, oxygen) or enzymatic causes. Oils and fats of

Table 1.3 Food antioxidants, maximum daily amounts and related possible toxicity

Additive code	Chemical name	Maximum daily dose	Possible toxicity
E 300	L-ascorbic acid	–	–
E 307	α-tocopherol synthesis	–	–
E 310 E 313	Gallates	0.1% in chewing gum 0.003% in potato flakes 0.01% in fats and oils (excluding olive oil) and potato flakes	Suspect sterility (in rats)
E 320 E 321	Butyl hydroxyanisole (BHA) Butyl hydroxytoluol (BHT)	0.1% in chewing gum 0.003% in flour and potato flakes 0.03% in French fries	Suspect renal complications
E 330	Citric acid	–	–

vegetable origin may resist without the addition of chemical stabilisers; however, certain oils and fats should be stabilised.

In general, shelf life extension requires the use of additives such as lecithins, tocopherols and polyphenols. The chemical browning (named Maillard reaction) can be counteracted by means of pH management, with the use of steel-made surfaces and in the absence of light and air (where possible). Enzymatic browning can be prevented or delayed with the use of antioxidants such as ascorbic acid, but much more effective is the inactivation of responsible enzymes (polyphenol oxidase, etc.). Oxidative rancidity, which mostly affects unsaturated fatty acids, is a result of oxygen (O_2) absorption and is favoured by light exposure and high temperatures. Chemically, this type of degradation will produce free radicals, peroxides and volatile compounds responsible for the characteristic rancid odour.

In general, antioxidants are defined primary or synergist antioxidants. Primary antioxidants are oxidised and important food components remain substantially unchanged, depending on the addition of chemicals and peculiar storage conditions. On the other side, synergists can reinforce the action of primary antioxidants (EFSA ANS 2011). The primary characteristic of oxidants is to fix O_2 and thus to prevent the oxidation of food components. Moreover, the antioxidant capacity is regenerated by the action of synergists by means of reduction reactions. Among the primary antioxidants, the following compounds can be mentioned: gallates, butylated hydroxyanisole (BHA), butylated hydroxytoluene (BHT), tocopherols and ascorbic acid. Secondary antioxidants, also known as synergists, are commonly found in food and considered innocuous when speaking of toxic effects (tartaric, lactic or citric acids, etc.). Table 1.3 shows several data with relation to toxicological data.

1.3.2.1 Chemical Classification

Chemically, antioxidants can be defined:

(1) Natural antioxidants. This category includes vitamins C and E (ascorbic acid and tocopherol respectively). These compounds are considered safe in foods; moreover, they have anti-cancer effects and inhibit oxidation reactions
(2) Natural–identical antioxidants. Chemically synthesised with the aim of copying the formula of natural ones, these substances are economically cheaper. An useful example is the chemical synthesis of ascorbic acid
(3) Synthetic antioxidants. These molecules do not exist in nature, and are the most contested and debated, even though they are approved at European level (examples: BHA and BHT).

Natural Antioxidants. Ascorbic Acid

It occurs naturally in many fruits such as citrus fruits, currants, blackberries, blueberries, elderberries, kiwi and raspberry. The natural identical L-ascorbic acid is produced by chemical synthesis for food industry purposes. Its broad use is due to the antagonistic effects against the reaction that originates the N-alkyl-nitrosamines.

It is found in various foods: jams and jellies, dry milk, fruit and frozen vegetables, pre-packed preparations of minced meat and beer.

Esters of ascorbic acid (coded as E 304) are generally added to fats (e.g. vegetable margarines) and oils with the exclusion of virgin oils and olive oil, because of their lipophilicity. The purpose is the prevention in oils, particularly those rich in polyunsaturated fatty acids. These esters may be also used as food colourants (E 160 and E 161) against the oxidation of pigments. Anyway, ascorbic acid and its derivatives do not produce any adverse effects in allowed concentrations for food production purposes (Mariani and Testa 2009).

Synthetic Antioxidants. BHA and BHT

These compounds have a very low cost if compared with other antioxidants. They are characterised by their activity at low concentrations (0.01–0.03% on fat weight); in addition, they may remain active after baking or frying treatments on fat molecules which have been added to. However, BHA is banned in many countries and in baby food because of carcinogenicity in liver. Some research conducted on rats showed that BHA leads to the formation of hydroxy radicals without possible detoxification if the product is taken in high doses. However, dose–response studies are needed to assess the security risk for humans (EFSA ANS 2011, 2012).

The European Parliament, following the positive opinion of the EFSA, has included in May 2015 the tertiary butyl hydroquinone (E 319, TBHQ) in the list of new antioxidant in fats and oils at a maximum of 200 mg/kg. TBHQ has a good solubility in fats and oils in foods, it is tasteless, it does not cause discoloration in the presence of iron. Finally, it can be used in combination with other antioxidants.

1.3.2.2 Other Antioxidants

Lecithins and polyphosphates, extracts mostly from soy, may be good antioxidants because of the presence of unsaturated bonds. Another antioxidant is ethylenedi-aminetetraacetic acid (EDTA), a synergist antioxidant with the ability to bind metals. Consequently, EDTA addition may allow the preservation of foods against oxidative browning. A complete antioxidant formula should contain two primary antioxidants, a synergist acid, a chelating agent for trace metals and a vehicle to disperse the above-mentioned components (Mariani and Testa 2009).

1.3.3 Structural Additives

Structural additives are related to the structure and sensory parameters of the food. Thickeners, stabilisers, gelling and wetting agents, and emulsifiers may act on the structural characteristics and appearance of foods.

Product details and food texture are the first parameters that consumers perceive when purchasing a food. Functional characteristics of foods affect all stages of food processing until storage in relation to their stability. These reflections are surely verified when speaking of baked goods, cheeses, milk-based desserts, puddings and canned meat. For these reasons, certain sensorial examinations concerning texture analysis—penetration, compression, shear, extrusion, tension, fracture and adhesiveness—are required (Arendt et al. 2007; Collar et al. 2005).

Structural additives are needed in the modern food industry if the desired texture and appearance of products is a key factor. Functional properties of these chemicals concern: viscosity, ability to form gels, foams stabilisation and solid suspensions. The choice of the 'right' additive is a function of required properties, temperature and pH. Although, these additives are foreign substances for human consumption, they are relatively safe by the toxicological viewpoint. For these reasons many substances have unlimited 'acceptable daily intake' (ADI) values. They are listed from E 400 to E 585. An EEC Directive has established structural additives according to their function: proper stabilisers, emulsifiers, gelling agents, turbidity, humectants and thickeners. All these substances have multiple properties and can be used with different purposes (Mariani and Testa).

1.3.3.1 Thickeners, Gelling Agents and Stabilisers

These additives, both natural and synthetic compounds, can give or ameliorate (and preserve) the desired consistency to the food product.

Natural substances in these categories are toxicologically safe and include:

- Pectin (polysaccharide found in fruits), used in puddings and jellies
- Alginates and agar-agar (extracted from seaweed), used for bakery
- Carob and guar (extracted from legume seeds), used for jellies and jams
- Carrageena, a mixture of anionic polymers, may increase food viscosity
- Polyphosphates (sodium, calcium and potassium salts). These compounds, widely used in the food industry (chewing gum, dried products, snacks and chocolate drinks), are able to bind calcium and magnesium ions. Moreover, they can solubilise actin and myosin (myofibrillar proteins), improve the emulsifying ability and the connection between different food portions (EFSA AFC 2006)
- Gelatin, obtained by the partial hydrolysis of connective tissues of animal origin. At present, the following additives are suspended or restricted in the EU: E 425 (*Konjac*) in jelly mini cups and confectionery (Directive of the European Community 2003/52/EC); gel-forming elements derived from seaweed and certain gums for choking hazard (jelly mini cups); E 400–E 407, E 407a, E 410, and E 412–E 415, E 417 and E 418 for jelly mini cups (Mariani and Testa 2009).

Stabilisers are used with the following results:

- Prevention of ice macroscopic crystals in ice creams (sorbitol, also used as sweetener, plays a recognised stabilising action)
- Prevention of structural collapses in dehydrated foods
- Creation of stable suspensions in beverages (precipitation phenomena and agglomeration are avoided)
- Prevention of sugar crystal dispersions in jams.

1.3.3.2 Emulsifiers

Emulsifiers have different properties, but their main feature concerns the capability of achieving and maintaining a uniform and stable dispersion of two immiscible phases.

Emulsifiers are classified as structural additives. These chemicals are structurally composed of a hydrophobic portion and a hydrophilic section; for this basic reason, the surface tension of a heterogeneous water/lipid system can be lowered. With relation to the chemical structure, they can be distinguished in: anionic, cationic, amphoteric and non-loaded substances. They are used for the production of margarines, ice cream products, spreads and mayonnaise. With reference to their origin, emulsifiers may be considered natural and synthetic emulsifiers.

Lecithins

Natural and synthetic lecithins are not completely acceptable as a single group of food additives. Some toxicological data with relation to lecithins and other emulsifiers are shown in Table 1.4. These emulsifying agents have a very long use; however, it is known that a good emulsion can be obtained with adequate homogenisation and the addition of appropriate egg amounts. In fact, eggs—as well as the albumen component lysozyme (E 1105)—contain natural emulsifiers such as lecithins and cholesterol.

Lecithins, chemically described as phospholipids by a polar head and an apolar tail, are the best emulsifiers for food production. In fact, these molecules can act as emulsifiers, stabilisers, anticaking agents and dietary supplements. In addition, they are reported to prevent fatty liver and increase levels of choline and acetylcholine in the blood and the human brain.

1.3.3.3 Humectants

Humectants are used in the food industry because of their ability to lose or absorb moisture. These chemicals have a great influence in the stability of 'intermediate moisture' (IMF) foods. The term IMF is used to indicate products with high percentage of free water. In general, the function of humectants can also be carried out by starches with unlimited ADI values, provided that these compounds are used following good practice rules (Mariani and Testa 2009). An important group of these substances are modified starches: these chemicals are obtained by chemical, physical or enzymatic hydrolysis with the aim of producing products with ameliorated water solubility. Toxicologists have doubts on the safety of metabolite derivatives for some of these starches (EFSA AFC 2006).

1.3.3.4 Flavour Enhancers

Flavour enhancers (classification codes: E 620–E 640) are substances commonly used in the food industry to enhance and intensify food flavour and aroma. Flavour

Table 1.4 Structural additives for food production purposes and related toxicity

Code	Chemical name	Relative toxicity
E 322	Lecithin	–
E 401	Sodium alginate	–
E 450	Disodium pyrophosphate	Observed hypocalcemia, accumulation of phosphate in the kidney, renal lesions associated with massive intake
E 450	Sodium polyphosphates	As above mentioned

enhancers have little or no flavour: however, they have the capacity to reinforce the natural flavour of foods. These food additives include: glutamates (examples: monosodium and monopotassium glutamates, E 621 and E 622 respectively), nucleotides and yeast extracts (Mariani and Testa 2009).

1.3.3.5 Dyes

A food colourant is a chemical compound (of organic or inorganic nature) that can be used with the aim of modifying the colorimetric appearance of a food product. Some dyes are natural substances, while others are synthetic imitations of natural compounds; certain chemicals may be modifications of original natural.

Chemically, dyes can be distinguished as follows:

- Natural dyes. These chemicals, toxicologically harmless and allowed for food uses, are generally derived from plants, through different extraction methods. However, natural dyes may be thermally unstable and sensible to pH and light; in addition, they are quite expensive and gladly accepted by the consumer
- Natural–identical dyes. These compounds are synthetic 'copies' of natural compounds
- Synthetic Dyes. These colours do not exist in nature. Synthetic dyes are surely cheaper than natural substances; on the other side, their appreciation by food consumers has decreases in recent years. In addition, these compounds are more stable to light, pH variations, oxidative reactions and related colours appear brighter and more uniform than natural substances. With relation to the chemical structure, synthetic dyes include azo and xantenic dyes, indigoids and anthraquinone.

Food dyes are classified in the 'range' E 100–E 199. In detail, natural and synthetic dyes are comprised between E 100 and E 163, while mineral pigments are between E 170 and E 180. Moreover, these substances can be distinguished depending on the use: mass and surface colouring, and exclusive superficial use (European Parliament and Council 1994a). Some exclusion has to be noted when speaking of vegetable extracts and other pigments (Scotter 2011).

The EFSA has clarified in 2002 that dyes can be added to food for some specific purposes:

(1) To offset the loss of colour due to exposure to air, moisture and light and temperature variations
(2) To enhance the natural colours and to add colour to foods that would otherwise be lacking or coloured differently.

EFSA's commitment in the field of food colouring uses aims to make clear safety assessments of new food colourants before they are authorised in the EU. In addition, all approved food additives have to be re-evaluated by the EU. As an example, two colours—Red 2G (E 128) and Brown FK (E 154)—have been removed from the list of allowed colourants in 2011 (Scotter 2011).

Natural Dyes

Dyes are obtained from plants or from certain animal species. The variety of colours that can be achieved are limited to red, yellow, blue and brown, while other colours can be obtained by mixing the four primary dyes.

The most used natural colours are:

(a) Yellow colours: curcumin (E 100, class of polyphenols; riboflavin (E 101): also called vitamin B_2;
(b) Red colours: cochineal (E 120); red *Orciello*, whose constituent is orcein (E 121)
(c) Green colours: chlorophyll (E 140), class of porphyrin pigments. The replacement of magnesium by chlorophyll degradation may turn the green colour into brown tints. A modified chlorophyll (E 141, copper complex) can be obtained easily
(d) Brown colours: caramel (E 150); this colour may be obtained easily from pale yellow to amber and dark brown tints by heating carbohydrates at specified temperatures
(e) Black colours: charcoal black (E 153). This substance is not easily solubilised in water; consequently, its use is very limited.

Other natural pigments in various shades are:

• Carotenoids. These substances—xanthophylls and carotenes—can be found in plants, algae and even some species of bacteria. There are more than 600 kinds of carotenoids. Carotenes (E 160) contain only carbon and hydrogen atoms; this category includes lycopene (tomatoes, peppers and apricots) and β-carotene which gives its name to the entire class present in carrots (orange colour). Xanthophylls (E 161) are naturally present in egg yolk, grapes plug and mandarins. This category includes two important dyes: lutein and zeaxanthin
• Betaine (E 162) is a dye used for candies, meat and ice creams The intensity of the colour depends on pH values (dark red between 4 and 5, blue-violet for pH > 5)
• Anthocyanins are a family of dyes belonging to flavonoids. By the chemical viewpoint, these polyphenolic structures are found in the flowers and fruits of almost all higher plants, and in the following foods: eggplant, berries, cherries and wines. Colours can range from red to blue tints; pH can influence the colorimetric appearance. Moreover, anthocyanins can react with molecular oxygen and free radicals by counteracting damages caused by these substances. They are used for jams and yoghurt in the food industry.

Synthetic Dyes

Synthetic colours are the most commonly used dyes at present because of their stability and low costs. On the other side, they may be associated with toxic effects (mainly because of the presence of azo groups in azo dyes and correlated genotoxic and carcinogenic effects). Consequently, current azo dyes for food production purposes contain have a sulfonic group which makes them soluble (easy removal) and without toxic effects (Combes and Haveland-Smith 1982).

The most known synthetic dyes are:

- Tartrazine (E 102). Used for candy, syrups and ice cream. It may sometimes be used to adulterate saffron
- Sunset Yellow (E 110). Uses for apricot jams, puddings, jellies, cheese and marzipan
- Red amaranth (E 123). Used for cocktail drinks and fish roe. It is not considered genotoxic (does not damage the genetic material of cells) or carcinogenic. The related European ADI is 0.15 mg per kg body weight per day; children's exposure was estimated to be around 30 times lower than ADI; consequently, it can be recommended in products intended for children (EFSA 2010)
- Allura Red (E 129). Similarly to E 123, it is an azo dye authorised as a food additive in the EU; it is also recommended in products intended for children. The current ADI is 7 mg/kg body weight per day
- Erythrosine (E 127). Used for the production of candies
- Indigo carmine (E 132) and Brilliant Blue (E 133). These colours are used for candies, candied fruit and cookies. The bright blue is also used in liquours
- Brilliant Black BN (E 151). This synthetic bis-azo dye is authorised as a food additive in the EU for use in foods in accordance with Regulation (EC) No 1333/2008 of the European Parliament and of the Council on food additives and subsequent amendments. In 2010, the EFSA Panel on Food Additives and Nutrient Sources Added to Food (ANS) adopted a scientific opinion on the re-evaluation of E 151 and concluded that dietary exposure in 1- to 10-year-old children at the high level may exceed ADI of 5 mg/kg body weight per day at the upper end of the range. Consequently, the European Commission requested that EFSA performs a refined exposure assessment for this food colour (EFSA 2015)
- Dark chocolate (155). It is prohibited in some European Countries, while it is allowed in the United States of America (USA) and Australia for chocolate desserts and flavoured milk chocolate.
- Bluish green (E 143). It is banned in Europe, but used in the USA for canned peas and ice cream.

1.3.4 Artificial Sweeteners

There are only a few tens of sweeteners used in the food industry. Basically, they can be subdivided into natural sweeteners and artificial sweeteners.

The aim of these chemicals concerns the necessity of a sweet taste in foods and beverages; in addition, they can be used as substitutes for sugar in some light products (soft drinks, nectars, chewing gum, some sweets, beer, yoghurt, etc.). However, different sweeteners have very different sweetness and caloric intake: only a few of these molecules can be considered the light version of sugar (sucrose).

In general, sweeteners are acceptable for people who cannot consume sugar (European Parliament and Council 1994b), although they are not recommended until the third year of age and during pregnancy and lactation, according to the Italian National Research Institute for Food and Nutrition (INRAN 2003) Particular attention should be paid to children over 3 years; the consumption of products containing sweeteners should be evaluated with caution.

Sweeteners are food additives which are used to impart a sweet taste in food-stuffs; in addition, table-top sweeteners may be considered. Sweeteners are regulated similarly to other food additives in the EU; the Commission and Member States decide which additives can be used in foods and at what levels. All food additives are included in the ingredient lists on product labels which must identify both the function of food additives in the finished food (i.e. sweetener) and the specific substance used either by referring to the appropriate E number or its name (e.g. E 954 for 'Saccharin') (EFSA 2013a).

1.3.4.1 Aspartame

Aspartame, N-L-α-aspartyl-L-phenylalanine 1-methyl ester, is a low-calorie, intense artificial sweetener. It is a white, odourless powder, approximately 200 times sweeter than sugar. It is authorised in the EU as a food additive in foodstuffs such as drinks, desserts, sweets, dairy, chewing gums, energy-reducing and weight control products and as a table-top sweetener. It has to be noted that EU labels on foodstuffs containing aspartame must declare its presence, indicating either its name or its E number (E 951). The main restrictions on use are linked to its instability to heat and pH. Aspartame rarely has side effects in sensitive people: headaches, nausea, vomiting and abdominal pain (EFSA 2013a). Moreover, it is not recommended for use during pregnancy because under certain conditions it dissociates into methanol and dipeptide which in turn can hydrolyse and cyclise giving rise to a compound toxic to the embryo. The use of aspartame is declared on labels with the 'Phenylketonurics' statement because it is a source of phenylalanine.

At present, aspartame and its breakdown products are considered safe for general population (including infants, children and pregnant women) by the EFSA. The current ADI of 40 mg/kg body weight per day is considered protective for the general population, and consumers' exposure to aspartame is well below this ADI.

However, this ADI is not applicable in patients suffering from phenylketonuria, as they require strict adherence to a diet low in phenylalanine (EFSA 2013b).

1.3.4.2   Advantame

Advantame is an intense artificial sweetener derived by chemical synthesis from isovanillin and aspartame. Chemical properties are different than those of aspartame. The sweetness of advantame can be hundreds or even thousands of times greater than that of sugar or other intense sweeteners, depending on the peculiar use.

At present, EFSA's experts concluded that advantame and its metabolites are neither genotoxic nor carcinogenic and pose no safety concern for consumers at the proposed uses and use levels as a sweetener. The current ADI is 5 mg/kg body weight per day; high levels of consumption in adults and children are significantly below this ADI. However, EU decision-makers have to decide whether advantame may be allowed as a sweetener in the EU.

1.3.4.3   Steviol Glycosides

Steviol glycosides (E 960) are mixtures of steviol glycosides used as sweetener; they can be extracted from leaves of the stevia plant. This sweetener has up to 300 times the sweetness of sugar but an almost negligible effect on blood glucose levels; hence it may be considered as an attractive substitute for sugar. In a scientific opinion published in April 2010, ANS experts concluded that steviol glycosides are neither genotoxic nor carcinogenic and established an ADI of 4 mg/kg body weight per day. On the other side, it was noted that this ADI could be exceeded by both adults and children if this sweetener is used at maximum levels proposed by the applicants. In fact, the EFSA reviewed in 2011 its previous assessment of consumer exposure based on revised levels of use proposed by the applicants. Adults and children who are high consumers of foods containing steviol glycosides could still exceed the ADI if the sweetener is used at maximum proposed levels. As a result, the EU Commission authorised in November 2011 with the Regulation (EU) N. 1131/2011 the use of steviol glycosides as sweeteners in foods with assigned number 'E 960' and addition to the official EU list of authorised food additives (European Commission 2011).

References

Arendt EK, Ryan LA, Dal Bello F (2007) Impact of sourdough on the texture of bread. Food Microbiol 24(2):165–174. Epub 2006 Sep 20. 10.1016/j.fm.2006.07.011

Collar C, Bollaín C, Angioloni A (2005) Significance of microbial transglutaminase on the sensory, mechanical and crumb grain pattern of enzyme supplemented fresh panbreads. J Food Eng 70(4):479–488. doi:10.1016/j.jfoodeng.2004.10.047

Combes RD, Haveland-Smith RB (1982) A review of the genotoxicity of food, drug and cosmetic colours and other azo, triphenylmethane and xanthene dyes. Mutat Res/Rev Genet Toxicol 98 (2):101–243. doi:10.1016/0165-1110(82)90015-X

Council of the European Communities (1988) Council Directive 89/107/EEC of 21 December 1988 on the approximation of the laws of the Member States concerning food additives authorized for use in foodstuffs intended for human consumption. Off J Eur Comm L 40:27–33

EFSA (2010) EFSA lowers ADI on amaranth, completing its re-evalution of azo dye food. European Food Safety Authority, Parma, Available https://www.efsa.europa.eu/en/press/news/ans100726. Accessed 16 Dec 2016

EFSA (2013a) EFSA explains the Safety of Aspartame. European Food Safety Authority, Parma, Available https://www.efsa.europa.eu/en/topics/factsheets/factsheetaspartame. Accessed 16 Dec 2016

EFSA (2013b) EFSA completes full risk assessment on Aspartame and Concludes it is safe at current level of exposure. European Food Safety Authority, Parma, Available https://www.efsa.europa.eu/en/press/news/131210. Accessed 16 Dec 2016

EFSA (2015) EFSA journal refined exposure assessment for brilliant black BN (E151). EFSA J 13 (1):3960–3993. doi:10.2903/j.efsa.2015.3960

EFSA AFC (2006) Opinion of the Scientific Panel on food additives, flavourings, processing aids and materials in contact whit food (AFC) related to the use of nisin (E234) as a food additives. EFSA J 4,3:314. doi:10.2903/j.efsa.2006.314

EFSA ANS (2011) Scientific opinion on the re-evaluation of butylated hydroxyanisole- BHA (E320) as a food additive. EFSA J 9, 10:2393. doi:10.2903/j.efsa.2011.2392

EFSA ANS (2012) Scientific opinion on the re-evaluation of butylated hydroxytoluene - BHT (E321) as a food additive. EFSA J 10, 3:2588. doi:10.2903/j.efsa.2012.2588

EFSA ANS (2014) Scientific Opinion on the re-evaluation of hexamethylene tetramine (E 239) as a food additive. EFSA J 12, 6:3696. doi:10.2903/j.efsa.2014.3696

EFSA CEF (2011) Scientific opinion on Flavouring Group Evaluation 20, Revision 3 (FGE.20Rev3): Benzyl alcohols, benzaldehydes, a related acetal, benzoic acids, and related esters from chemical groups 23 and 30. EFSA J 9,7(2176):136. doi:10.2903/j.efsa.2011.2176

European Commission (2011) Commission Regulation (EU) No 1131/2011 amending Annex II to Regulation (EC) No 1333/2008 of the European Parliament and of the Council with regard to steviol glycosides. Off J Eur Union L 295:205–211

European Commission (2015) Commission Regulation (EU) 2015/1725 of 28 September 2015 amending Annex to Regulation (EU) No 231/2012 laying down specifications for food additives listed in Annexes II and III to Regulation (EC) No 1333/2008 of the European Parliament and of the Council as regards specifications for Ethyl lauroyl arginate (E 243). J Off J Eur Union L 252:12–13

European Parliament and Council (1994a) European Parliament and Council Directive 94/36/EC of 30 June 1994 on colours for use in foodstuffs. Off J Eur Comm L 237:13–29

European Parliament and Council (1994b) European Parliament and Council Directive 94/35/EC of 30 June 1994 on sweeteners for use in foodstuffs. Off J Eur Comm L 237:3–12

European Parliament and Council (2010) Commission Regulation (EU) No 257/2010 of 25 March 2010 setting up a program for the re-evaluation of approved food additives in accordance with Regulation (EC) No 1333/2008 of the European Parliament and of the Council on food additives. Off J Eur Union L 80:19–27

Garcia-Fuentes AR, Wirtz S, Vos E, Verhagen H (2015) Short review of sulphites as food additives. Eur J Nutr Food Saf 5(2):113–120. doi:10.9734/EJNFS/2015/11557

Hawkins DR, Rocabayera X, Ruckman S, Segret R, Shaw D (2009) Metabolism and pharmacokinetics of ethyl N(alpha)-lauroyl-L-arginate hydrochloride in human volunteers. Food Chem Toxicol 47(11):2711–2715. doi:10.1016/j.fct.2009.07.028

Honikel KO (2008) The use and control of nitrate and nitrite for the processing of meat products. Meat Sci 78(1–2):68–76. doi:10.1016/j.meatsci.2007.05.030

INRAN (2003) Linee Guida per una sana alimentazione italiana. Ministero delle Politiche Agricole e Forestali, Rome, and Istituto Nazionale di Ricerca per gli Alimenti e la Nutrizione, Rome

Mariani M, Testa S (2009) Gli Additivi alimentari. Macro Edizioni, Diegaro di CesenaS

Ministero della Salute (2012) Lista Comunitaria degli additivi autorizzati negli alimenti di origine vegetale e animale. Ministero della Salute, nota n. 21863 DGISAN - 6/I.4.c.c.8.7/2. n. 21863 – P 18/06/2012. Available http://www.salute.gov.it/portale/temi/p2_6.jsp?lingua=italiano&id=4450&area=sicurezzaAlimentare&menu=additivi. Accessed 14 Dec 2016

Ruckman SA, Rocabayera X, Borzelleca JF, Sandusky CB (2004) Toxicological and metabolic investigations of the safety of N-alpha-lauroyl-L-arginine ethyl ester monohydrochloride (LAE). Food Chem Toxicol 42(2):245–259. doi:10.1016/j.fct.2003.08.022

Saltmarsh M (ed) (2000) Essential guide to food additives. Leatherhead Publishing, LFRA Ltd., Leatherhead

Scotter MJ (2011) Emerging and persistent issues with artificial food colours: natural colour additives as alternatives to synthetic colours in food and drink. Qual Assur Saf Crop Food 3 (1):28–39. doi:10.1111/j.1757-837X.2010.00087.x

Chapter 2
The Codex Alimentarius and the European Legislation on Food Additives

Pasqualina Laganà, Emanuela Avventuroso, Giovanni Romano,
Maria Eufemia Gioffré, Paolo Patanè, Salvatore Parisi,
Umberto Moscato and Santi Delia

Abstract The aim of this Chapter is to give an overview of International principles concerning food additives with a description of the related European legislation. Basically, the matter of food additives may be discussed on a large-scale level by means of the description of the 'General Standards for Food Additives' (Codex Alimentarius Commission), a harmonised, workable and indisputable international standard. On the other hand, the European viewpoint has to be considered: main Regulations concerning food additives have been discussed. The EU Legislation and Codex Alimentarius have very similar regulations about food additives, sometimes using the same definitions. This situation shows a general trend in the international harmonisation of technical legislation in many law fields. However, some difference has to be highlighted, including the legal validity. In fact, Codex Alimentarius documents have established a sort of general principle that has to be adopted by domestic laws to become effective. In contrast, European Union Regulations are legally mandatory for all Member states of the Union after their official publication and national translation.

Keywords Acceptable daily intake · Carry-over principle · Community list · European union · Good manufacturing practices · International numbering system

Abbreviations

ADI	Acceptable daily intake
CAC/GL 36-1989	Class Names and the International Numbering System for Food Additives
CODEX STAN 107-1981	Codex Standard 107-1981
CODEX STAN 192-1995	General standard for food additives
EU	European Union
FAO	Food and Agriculture Organization of the United Nations
INS	International Numbering System
WHO	World Health Organization

© The Author(s) 2017
P. Laganà et al., *Chemistry and Hygiene of Food Additives*,
Chemistry of Foods, DOI 10.1007/978-3-319-57042-6_2

2.1   Introduction to the Codex Alimentarius and the European Legislation on Food Additives

The aim of the present chapter is to give an overview of International principles concerning food additives with a description of the related European legislation.

2.1.1   The Codex Alimentarius. A Definition

The Codex Alimentarius can be defined as a collection of standards, codes of practice, guidelines and recommendations on foods issued by a joint Commission between the Food and Agriculture Organization of the United Nations (FAO) and the World Health Organisation (WHO). The Codex Alimentarius Commission, established in the 1960s by the two said organisations, is the most important International reference point for food standards. The aim of the Codex is to create food standards in order to protect the consumer and, at the same time, facilitate trade and ensure fair practices in trade by the help of all involved stakeholders (Codex Alimentarius 2006).

2.1.2   Food Additives. The Codex Alimentarius Viewpoint

The General standard for food additives (CODEX STAN 192-1995) of Codex Alimentarius, last reviewed in 2015, states that Food Additive means '*any substance not normally consumed as food by itself and not normally used as a typical ingredient of the food, whether or not it has nutritive value, the intentional addition of which to food for a technological (including organoleptic) purpose in the manufacture, processing, preparation, treatment, packing, packaging, transport or holding of such food results, or may be reasonably expected to result (directly or indirectly), in it or its by-products becoming a component of or otherwise affecting the characteristics of such foods. The terms do not include contaminants or substances added to food for maintaining or improving nutritional qualities*'. Consequently, food additives are used for a technological purpose in the process of food production.

With relation to safety, food additives have to comply with the following requests, in accordance with CODEX STAN 192-1995:

(a) No appreciable health risks to consumers are shown when speaking of endorsed food additives (it has to be considered that the use implies proposed use levels)
(b) Endorsed food additives are always associated with an acceptable daily intake (ADI) value or similar safety assessment; in addition, the probable daily intake has to be carefully considered when speaking of special consumer groups

(c) The intended technical effect for the use of endorsed chemicals determines the lowest amount of these additives in foods.

Moreover, the **CODEX** STAN 192-1995 does not give full freedom with reference to the use food additives (Codex Alimentarius 1995). In fact, the following requests have to be satisfied:

(a) The addition of these chemical substances has to imply an advantage
(b) Consumers cannot be damaged; no misleading has to be caused
(c) One or more of the technological functions and similar justifications set out in CODEX STAN 192-1995, paragraph 3.2, letters from (a) to (d), have to be satisfied.

Moreover, 'Good Manufacturing Practices' have to be considered when speaking of food additives. The amount of added substance(s) has to be related with the desired advantage and the lowest addition is mandatory. The same concept is repeated when considering the minimum absorption of a peculiar food additive by foods as the result of its use in manufacturing, processing or packaging steps. Finally, each food-grade additive has to show good qualitative features and be considered as a food ingredient.

2.2   Classification of Food Additives: The International Numbering System

The 'Class Names and the International Numbering System for Food Additives' (CAC/GL 36-1989), first adopted in 1989 and last amended in 2016, has created the International Numbering System (INS) for Food Additives: a system which aims to provide a harmonised naming system for food additives (Codex Alimentarius 1989). However, INS inclusion does not mean that the Codex Alimentarius approves included food additives. The INS may also consider those additives that have not been evaluated by the joint FAO/WHO Expert Committee on Food Additives. In general, the system corresponds to an open list subject to the inclusion of additional additives or the removal of existing ones on an ongoing basis. The list of functional classes of additives (European Commission 2011) with one or more examples for every class is briefly described as follows, while the functions of several additives are explained in Chap. 1:

- Acidity regulator (i.e. *calcium carbonate*)
- Anticaking agent (i.e. *synthetic magnesium silicate*)
- Antifoaming agent (i.e. *microcrystalline wax*)
- Antioxidant (i.e. phosphoric acid, potassium ascorbate)
- Bleaching agent (i.e. sodium sulfite, sulphur dioxide, chlorine)
- Bulking agent (i.e. isomalt or hydrogenated isomaltulose)
- Carbonating agent (i.e. carbon dioxide)

- Carrier (i.e. calcium lactobionate, potassium alginate)
- Colour (i.e. chlorophylls, curcumin, erythrosine)
- Colour retention agent (i.e. magnesium carbonate, nicotinic acid)
- Emulsifier (i.e. powdered cellulose)
- Emulsifying salt (i.e. sodium and calcium lactate)
- Firming agent (i.e. dimagnesium diphosphate, potassium chloride)
- Flavour enhancer (i.e. monosodium succinate, magnesium gluconate)
- Flour treatment agent (i.e. calcium and potassium iodate)
- Foaming agent (i.e. nitrogen, nitrous oxide, ammonium alginate)
- Gelling agent (i.e. carrageenan, *cassia* gum, *konjac* flour)
- Glazing agent (i.e. lanolin, methyl esters of fatty acids)
- Humectant (i.e. pentapotassium triphosphate, polydextroses)
- Packaging gas (i.e. carbon dioxide, nitrous dioxide)
- Preservative (i.e. lysozyme, benzoic acid, calcium benzoate)
- Propellant (i.e. isobutane, dichlorodifluormethane, butane)
- Raising agent (i.e. sodium dihydrogen phosphate, tricalcium phosphate, sodium polyphosphate)
- Sequestrant (i.e. ammonium polyphosphate)
- Stabiliser (i.e. sucrose esters of fatty acids, sorbitan trioleate)
- Sweetener (i.e. acesulfame potassium, aspartame, cyclamic acid)
- Thickener (i.e. magnesium hydrogen phosphate).

Some additives may have different uses. A useful example is calcium carbonate: it may act as acidity regulator, anticaking agent, surface colourant, firming agent, flour treatment agent (dough conditioner) and stabiliser. Another situation is shown when speaking Gum Arabic (or Acacia Gum) which can be used as bulking agent, carrier, emulsifier, glazing agent, stabiliser and thickener.

2.3 Food Additives and Labelling. The Codex Viewpoint

The Codex Standard 107-1981 (hereinafter CODEX STAN 107/1981) contains rules for the labelling of food additives. Article 2, letter f) of the said document gives the definition of 'label' as follows: '*label includes any tag, brand, mark, pictorial or other descriptive matter, written, painted, stencil led, marked, embossed or impressed on, or attached to, a container*' In addition, 'labelling' is defined (Article 2 letter e) in the following way: '*labelling includes the label and any written, printed or graphic matter relating to and accompanying the food additives. The term does not include bills, invoices and similar material which may accompany the food additives*' (Codex Alimentarius 1981).

With relation to general principles on food additives and labelling requirements, Articles 3 and 4 of Codex Stan 107/1981 give important information and instructions on mandatory labels of prepackaged food additives sold by retail. The interested reader is invited to consult this document in detail. By a general viewpoint, the language for labels must be acceptable with relation to the receiving Country. Should this language be unacceptable, two different options would be available:

(a) Re-labelling
(b) Addition of another label to the first label. The second label must contain mandatory information written in an acceptable language.

Written information on labels has to be clear, prominent and readily legible by the consumer under normal conditions of purchase and use. In addition, should food additives have been treated with ionising radiations, the label would necessarily mention the treatment.

2.4   Food Additives and the European Legislation

With relation to the European Union (EU), the definition of additives is mentioned in the Regulation (EC) No 1333/2008 of the European Parliament and of the Council of 16 December 2008 on food additives (European Parliament and Council 2008a). In particular, Article 3 paragraph 2, letter a) states that food additives '*shall mean any substance not normally consumed as a food in itself and not normally used as a characteristic ingredient of food, whether or not it has nutritive value, the intentional addition of which to food for a technological purpose in the manufacture, processing, preparation, treatment, packaging, transport or storage of food results, or may be reasonably expected to result, in it or its by-products becoming directly or indirectly a component of such foods*'. This definition is almost identical if compared with the analogous Codex Alimentarius description, showing the general trend of International legislation to the harmonisation.

Differently from the Codex approach, this EU Regulation n° 1333/2008 allows only food additives included in the Community list in Annexes II and III (European Parliament and Council 2008a). Mentioned food additives may be placed on the market and used under the conditions of use specified in the said Regulation as set in the provisions of Article 4. Article 5 states also that '*No person shall place on the market a food additive or any food in which such a food additive is present if the use of the food additive does not comply with this regulation*'.

Generally speaking, a food additive can be included in the Community list provided that several conditions are satisfied (Reg. n° 1333/2008, Article 6, paragraph 1, letters a, b, and c). With relation to the Codex Alimentarius viewpoint, some points should be highlighted because of observed differences:

(a) Food additives can be used on condition that the reasonable technological need cannot be obtained in other economically convenient or workable ways
(b) Food additives have to explicitly guarantee advantages to the consumer, according to Reg. n° 1333/2008, Article 6, point 2.

Moreover, the EU legislation aims to protect the health of the consumer and His ability to make discriminations between different products on the market.

In detail, specific conditions are set in Articles 7 and 8 for some products that are supposed to have additional specific requisites. Two examples are sweeteners and colours. Sweeteners are intended to:

- Replace sugars for the production of energy-reduced food
- Replace sugars in non-cariogenic foods
- Replace sugars in foods with no added sugars
- Replace sugars if the aim is the increase of food shelf life.

With relation to colours (Chap. 1), they are able to: Restore the original appearance of foods

- Give a more interesting appearance to foods if compared to the original coloured product
- Colorise colourless foods.

The European classification of food additives concerns 27 different groups in accordance with Reg. n° 1333/2008, Annex I. Two basic differences are observed if these classes are compared with CAC/GL 36/1989.

In detail, acids, contrast enhancers and modified stanches are included in Reg. n° 1333/2008 while CAC/GL 36/1989 does not mention them directly. An acid is defined by Reg. n° 1333/2008 as a substance which increases the acidity of a foodstuff and/or imparts sour tastes. On the other hand, contrast enhancers are defined as substances able to reveal colorimetric differences between pre-treated selected areas of external surfaces (fruits or vegetables) following the interaction with certain components of the epidermis. Finally, modified starches are described as substances obtained by one or more chemical treatments of edible starch.

On the contrary, CAC/GL 36/1989 lists some food additives that are not included in the said EU Regulation: bleaching, carbonating and colour retention agents.

Another important point is the definition of use levels for food additives: Reg. n° 1333/2008, Article 11, point 1 establishes the levels of use of food additives. Once more, use levels have to be the lowest needed amounts for the desired effect and have to be chosen taking into account ADI values and the possible daily intake of food additives by 'special' consumers or gr. In addition, Articles 15 and 16 establish specific prohibitions in some foods or especially for particular consumer categories of consumers. It has to be noted that these prohibitions may show many exceptions. An useful example is sodium alginate (E 401): in accordance with Reg. n° 1333/2008, Annex II, it can be used at the maximum level of 1,000 mg/l (or mg/kg where appropriate) for '*Dietary foods for babies and young children* for special medical purposes *from four months onwards in special food products with adapted composition, required for metabolic disorders and for general tube-feeding*' (European Parliament and Council 2008a). Definitions of 'infants' and 'young children' may be found in Directive 98/398/EEC.

One of the most important exceptions dealing with the authorised presence of a food additive is the so-called 'carry-over principle', settled down in Article 18, point 1, letter a of Reg. n° 1333/2008 (European Parliament and Council 2008a). This principle states that '*the presence of a food additive is permitted in a compound food where the food additive is permitted in one of the ingredients of the compound food*'. As a result, a specific food additive allowed in one food is also allowed in compound foods that uses the first food as its own ingredient. Several exceptions are notable, in accordance with Annex II, Table 1.1, part a, including unprocessed food, honey, non-emulsified oils and fats of animal and vegetable origin, butter, unflavoured pasteurised and sterilised (including UHT) milk and unflavoured plain pasteurised cream, unflavoured fermented milk products, natural mineral water, spring water and any other bottled or packed water, coffee and coffee extracts, unflavoured leaf tea, sugars and dry pasta (excluding gluten-free and/or pasta intended for hypoproteic diets).

2.5 Traditional Foods and Food Additives

The EU legislation, at of establishes (Reg. n° 1333/2008, Article 20) that some Member states may continue to prohibit the use of certain food additives or related categories when speaking of their traditional products. The list with 13 total exceptions is mentioned in the Annex IV of Reg. n° 1333/2008. So, for example:

(a) France can ban the use of every food additive from traditional products including French bread, traditional French preserved snails and traditional French goose and duck preserves (confit)
(b) German can continue to ban the use of all food additives except propellant gases in traditional German beer
(c) Italy can ban food additives except preservatives, antioxidants, pH-adjusting agents, flavour enhancers, stabilisers and packaging gas from traditional Italian Mortadella and traditional Italian Cotechino and Zampone
(d) Sweden and Finland can ban only colours from traditional Swedish and Finnish fruit syrups.

Apart these exceptions, the general use of food additives is allowed; on the other hand, this use depends on the production process, used ingredients, the final appearance and other requirements. In addition, allowed substances may be naturally found in unprocessed foods: for example, apples may contain riboflavins (E 101), carotenes (E 160a), anthocyanins (E 163), acetic acid (E 260), ascorbic acid (E 300), citric acid (E 330), tartaric acid (E 334), succinic acid (E 363), glutamic acid (E 620) and L-cysteine (E 920).

2.6   Safety Assessment of Food Additives in the EU

Briefly, the safety assessment and the correlated authorisation procedure of food additives, enzymes and flavourings are performed in the EU as stated by Regulation n° 1331/2008 (European Parliament and Council 2008b). The basic principle concerning the assessment and authorization procedure for these three categories states that the common procedure must be '*effective, time-limited and transparent*', so as to facilitate the free movement within the community market. However, the authorisation for food additives used or intended for use in or on foodstuffs must be preceded also by an assessment of health risks for human health. It should be highlighted that food additives may be used as ingredients or intended for external (superficial) purposes.

The procedure for authorization to place a food additive on the market can be summarised as follows (Reg. n° 1331/2008):

(a) The European Commission can start the procedure on his own initiative or following an application by a Member State or by an interested party. Applicants may be producers or potential users for this food additive, described with all relevant data for the assessment (chemical identification, manufacturing process, methods of analyses, the case of need, toxicological data and proposed use)

(b) The Commission can seek for the EFSA (time for opinions: 9 months, with possible exceptions). The EFSA opinion must be forwarded to the Commission, the Member State and, if there is any, the applicant

(c) The Commission is assisted in the procedure by the Standing Committee on the Food Chain and Animal Health. Within 9 months starting from the date of the opinion released by the EFSA, the Commission has to submit a draft regulation updating the Community list of allowed food additives on the basis of EFSA's opinions and other factors. Interestingly, the Commission may adopt discordant decisions if compared with EFSA's opinions, with clear justifications

(d) The Commission can update the list of allowed food additives (options: addition of a new allowed substance; removal; addition, removal or modification of conditions, restrictions or specifications)

(e) Finally, the updated list is published in the Official Journal of the European Union.

It has to be noted that the entire procedure has to assure transparency and confidential treatment of all relevant information (Regulation n° 1331/2008, Articles 11 and 12).

2.7   Food Additives in the EU and Emergency Measures

The Regulation (EC) N° 178/2002 establishes emergency measures for selected situations concerning foodstuffs and possible serious risks to human health, animal health or the environment. In particular, Article 53 of Reg. N° 178/2002 deals with emergency measures for food and feed of Community origin or imported from a third Country. The following measures have to be immediately adopted by the Commission (European Parliament and Council 2002):

– Suspension of the placing on the market or use of the suspected food(s) or feed(s)
– Laying down special conditions for the suspected food(s) or feed(s)
– Any other appropriate *interim* measure.

Similar measures can be adopted referring to foods or feeds imported from a third Country.

2.8   Conclusion

The EU Legislation and Codex Alimentarius have very similar regulations about food additives, sometimes using the same definitions. This shows a general trend in the international harmonisation of technical legislation in many law fields. The difference in the validity of those two different law systems has to be highlighted. In fact, Codex Alimentarius documents have established a sort of general principle that has to be adopted by domestic laws to become effective. On the other side, EU Regulations are legally mandatory for all Member states of the Union on the 20th day after their publication on the Official Journal of the EU. Moreover, EU Directives have to be adopted by Member states and translated into national laws within a term fixed by EU Institutions.

References

Codex Alimentarius (1989) Class names and the International numbering system for food additives CAC/GL 36-1989 adopted in 1989, revision: 2008, amendment: 2015. Food and Agriculture Organization of the United Nations, Rome, and World Health Organization, Geneva

Codex Alimentarius (1995) General standard for food additives CODEX STAN 192-1995, adopted in 1995, revision 2015. Food and Agriculture Organization of the United Nations, Rome, and World Health Organization, Geneva

Codex Alimentarius (1981) General standard for the labelling of food additives when sold as such, CODEX STAN 107-1981. Food and Agriculture Organization of the United Nations, Rome, and World Health Organization, Geneva

Codex Alimentarius (2006) Understanding the Codex Alimentarius, third edition. the Secretariat of
    the Joint FAO/WHO Food Standards Programme, FAO, Rome
European Parliament and Council (2002) Regulation (EC) n° 178/2002 of the European Parliament
    and of the Council of 28 January 2002 laying down the general principles and requirements of
    food law, establishing the European Food Safety Authority and laying down procedures in
    matters of food safety. Off J Eur Comm L 31:1–24
European Parliament and Council (2008a) Regulation (EC) No 1333/2008 of the European
    Parliament and of the Council of 16 December 2008 on food additives, lastly amended by
    Commission Regulation No 2015/538 of 31 March 2015. Off J Eur Comm L 354:16–33
European Parliament and Council (2008b) Regulation (EC) No 1331/2008 of the European
    Parliament and of the Council of 16 December 2008 establishing a common authorisation
    procedure for food additives, food enzymes and food flavourings. Off J Eur Comm L 354:1–6
European Commission (2011) Questions and answers on food additives commission Européenne
    —MEMO/11/783 14/11/2011. Available http://europa.eu/rapid/press-release_MEMO-11-783_
    en.htm?locale=FR. Accessed 16 Dec 2016

Chapter 3
Food Additives and Effects on the Microbial Ecology in Yoghurts

Pasqualina Laganà, Emanuela Avventuroso, Giovanni Romano,
Maria Eufemia Gioffré, Paolo Patanè, Salvatore Parisi,
Umberto Moscato and Santi Delia

Abstract The aim of this study was to describe the current use of selected food additives in the dairy sector, particularly with relation to the production of yoghurt products. The modern production of yoghurt is performed by means of controlled processes and the careful selection of ingredients in addition to bovine milk: milk powder, sugar, fruit and bacterial starter cultures, flavourings, colour agents, emulsifiers, and stabilisers. These substances are grouped under the name of food additives. The use of certain additives, especially sweeteners, dyes and thickening agents, has been discussed here with relation to peculiar yoghurt categories which can represent the current situation of the whole market in terms of global sales and consumers' preferences.

Keywords Dye · Enriched yoghurt · Lactose-fermenting bacteria · Plain yoghurt · Sweetener · Thickening agent · Yoghurt-type dessert

Abbreviations
Lab   Lactose-fermenting bacteria

3.1   Introduction to Food Additives in the Yoghurt Industry

The lifestyle and eating habits affect greatly the population's health status. Social and cultural changes of the modern society can be seen as one of the main causes for unhealthy diets and the increased consumption of certain ready-to-eat foods, easy to prepare and sometimes poor in nutrients. These factors may lead to a considerable increase of diseases, significantly impacting on the person's health conditions (Markowitz and Bengmark 2002). An excessive and inappropriate use of antibiotics was noted until a few years ago; at present, industries have focused their efforts with relation to probiotics, defined as 'live microbes which pass through the

© The Author(s) 2017   33
P. Laganà et al., *Chemistry and Hygiene of Food Additives*,
Chemistry of Foods, DOI 10.1007/978-3-319-57042-6_3

gastro-intestinal tract leading to the health benefits of the consumer' (Tannock et al. 2000).

The increased use of probiotics has also promoted the augment of food additives in certain products. The best-known dairy product with living probiotic bacteria such as *Lactobacillus acidophilus* and *Bifidobacterium bifidum* is certainly the yoghurt (Baglio 2014; Lourens-Hattingh and Viljoen 2001). This food is a product resulting from the heat treatment of milk with the needed action of lactose-fermenting bacteria such as *L. bulgaricus* and *Streptococcus thermophilus* (Chandan and Kilara 2013). The main role of lactic bacteria is to use lactose in milk as a substrate with the production of lactic acid and the consequent pH lowering (until 4.6) during milk fermentation; fermentation has to me managed properly (Daravingas et al. 2001; Kneifel et al. 1993; Laroia and Martin 1991; Lourens-Hattingh and Viljoen 2001; Modler et al. 1990; Tamime and Robinson 1985).

From the consumer's point of view, the success of yoghurts appears to be due to the slightly sour taste of fermented milk that makes it a fresh food and therefore particularly pleasing. From a commercial perspective, various yoghurts are available at present:

- Plain yoghurt produced by milk only
- Yoghurt-type desserts (ingredients: milk, juice or puree of fruits and other ingredients such as sugar, cereals, malt, chocolate, coffee, royal jelly, etc.)
- Enriched yoghurt (plain yoghurt or yoghurt-type desserts that are enriched with minerals, vitamins, fibre, or oligosaccharides).

Typically, yoghurt is produced from whole, skimmed milk, or partially skimmed milk, according to the desired fat content in the finished product. The basic ingredient is milk, without the presence of residues of antibiotics and detergents, because it is known the high sensitivity of lactose-fermenting bacteria (LAB). The presence of cleaning or antibiotic residues can slow lactic fermentation resulting in the insufficient acidification.

The modern production of yoghurt is performed by means of controlled processes and the careful selection of ingredients in addition to bovine milk: milk powder, sugar, fruit and bacterial starter cultures, flavourings, colour agents, emulsifiers, and stabilisers. These substances are grouped under the name of food additives (Chap. 2).

3.2 Yoghurt and Some Recommended Additives

Food additives correspond to a group of edible substances which are not consumed as foods as such, nor are used as typical ingredients of foods, but they are added to foodstuffs to perform particular technological functions (Sect. 2.1.2; Codex Alimentarius 1995). At present, the usefulness of additives in the modern food industry is mainly related to the need to obtain food or drink with specific tech-nological properties or sensory characteristics without other possible or economi-cally convenient strategies (Sects. 2.1.2 and 2.4).

By the European legal viewpoint, food additives have been discussed in Chap. 2 of this book (Sects. 2.4–2.7).

It should be considered that all allowed food additives are used on condition that the maximum level for use has no evident toxic effects on human beings. The determination of 'acceptable daily intake' (ADI) values for each food additive provides a large margin of safety and refers to the amount of food additive that can be usually added in the daily diet, even for the whole lifetime without risk.

Modern food technology and marketing strategies have increasingly proposed the use of food additives as an essential component of the yoghurt production. In other terms, modern yoghurts should show certain features and special properties such as increased viscosity, when necessary, enhanced sweetness and aroma, colour improvement, and ameliorated durability of the final product. These sensorial features should remain constant until the final day of durability periods: this necessity seems to justify the important role of modern food additives.

With exclusive relation to yoghurts, food additives have to be mentioned without possible contrasting effects when speaking of food labels.

For example, sweeteners or emulsifiers cannot be added and consequently mentioned on yoghurt labels if the same labels contain the words 'natural yoghurt'. On the one hand, commercial strategies often try to attract the consumer with specific advertising messages such as the enhanced 'creaminess' of certain prod-ucts. On the other side, the need of specific additives has to be carefully considered.

One of the more interesting yoghurt categories in this ambit (when speaking of notable additions) is surely the 'fruit yoghurt', probably because there are very little amounts of added fruits. As a result, industries may incorporate aromas able to simulate the desired taste, deceiving consumers to eat a fruit-rich product. Since real fruits are added in limited quantities, dyes are added to give a consistent visual appearance. The use of dyes (Sects. 1.3.3.5.1 and 1.3.4) is particularly important for products made from red fruits (cherries, strawberries, wild berries); otherwise, these products would not be reliable enough. Moreover, sweet yoghurts may mean also sugar addition, giving a very caloric intake. Alternatively, artificial sweeteners such as aspartame (E 951) and acesulfame (E 950) may be used for the production of 'light' yoghurt (Sect. 1.3.4) although natural sweeteners may be considered. Anyway, these additives are required to be as sweet as saccharose, non-cariogenic, colourless and odourless, non-toxic, and without negative effect for consumers' health (O'Brien Nabors 2011).

Since E 950 and E 951 have a more enhanced sweetening power if compared with sucrose while their caloric power is almost zero, the replacement of saccharose with these substances can significantly reduce caloric intake of foods. The use of synthetic sweeteners has often been criticised, because the sugar taken could have adverse consequences on the health of the consumer; for this reason natural sub-stances are recommended. As with almost all food additives, abuse can cause damages to the organism; for these reasons, recommended ADI should be always recommended.

As previously mentioned, dyes can be widely used when speaking of yoghurts. Currently allowed colour agents include synthetic and natural substances. The choice of the 'right' dye is made according to the yoghourt ingredients; caramel (E 150d) is recommended for coffee-flavoured yoghurt, while cochineal (E 120)— which is a natural red dye derived by the ladybug—is used for products with strawberry-, cherry- and berry- flavoured products. Anyway, selected dyes must be resistant to extreme conditions during production steps such as long heat treatments, an excessive amount of lipids (in the organic phase) or the migration of particles (Daravingas et al. 2001).

3.3   Food Additives for Yoghurt in Brief. The Use of Thickening Agents

At present, rheological properties of normal yoghurt and yoghurt drinks depend mainly on the addition of thickeners, emulsifiers and stabilisers (Sect. 1.3.3).

Thickening agents can be of both natural and artificial origins. The most used are the natural ones; normally, they are of vegetable origin, but thickeners of animal origin can also exist. Thickeners are often used to enhance low-quality products by means of their ability to absorb or bind water, with consequent increase of density and the creaminess of the finished product. However, thickeners may reduce the nutritional value; because of the augment of water molecules and the consequent decrease in the amount of vitamins and mineral salts, thickeners are widely used in the preparation of light yoghurt. The most used agents in this ambit include pectin (E 440), corn starch, guar gum (E 412), the xanthan gum (E 415) and alginates. Some of these substances have a dual function. For example, pectin or xanthan gum can also be defined as gelling agents because of the creation of colloidal structures (promotion of gelified networks in emulsified foods). An interesting application concerns alginates in yoghurt; these thickeners have been reported to increase the survival and viability of probiotic bacteria in yoghurt during storage.

References

Baglio E (2014) The industry of yoghurt: formulations and food additives. In: Baglio E (ed) Chemistry and technology of yoghurt fermentation. SpringerBriefs in Chemistry of Foods, Springer, Cham

Chandan RC, Kilara A (eds) (2013) Manufacturing yogurt and fermented milks, 2nd edn. Wiley, Chichester, pp 297–317

Codex Alimentarius (1995) General standard for food additives CODEX STAN 192-1995, adopted in 1995, revision 2015. Food and Agriculture Organization of the United Nations, Rome, and World Health Organization, Geneva

Daravingas GV, Heitke TC, Funk DF (2001) Colored multi-layered yogurt and methods of preparation. US Patent 6,235,320, 22 May 2001

Kneifel W, Jaros D, Erhard F (1993) Microflora and acidification properties of yogurt and yogurt-related products fermented with commercially available starter cultures. Int J Food Microbiol 18(3):179–189. doi:10.1016/0168-1605(93)90043-G

Laroia S, Martin JH (1991) Effect of pH on survival of *Bifidobacterium bifidum* and *Lactobacillus acidophilus* in frozen fermented dairy desserts. Cult Dairy Prod J 2:13–21

Lourens-Hattingh A, Viljoen BC (2001) Yogurt as a probiotic carrier. Int Dairy J 11(1–2):1–17. doi:10.1016/S0958-6946(01)00036-X

Markowitz JE, Bengmark S (2002) Probiotics in health and disease in the pediatric patient. J Pediatr Gastroenterol Nutr 49(1):127–141

Modler HW, McKellar RC, Yaguchi M (1990) Bifidobacteria and bifidogenic factors. Can Inst Food Sci Technol J 23:29–41

O'Brien Nabors L (2011) Alternative sweeteners: an overview. In: O'Brien Nabors L (ed) Alternative sweeteners, vol 48. CRC Press, Boca Raton

Tamime AY, Robinson RK (1985) Yogurt: science and technology. Pergamon Press, Oxford

Tannock GW, Munro K, Harmsen HJM, Welling GW, Smart J, Gopal PK (2000) Analysis of the fecal microflora of human subjects consuming a probiotic product containing *Lactobacillus rhamnosus* DR 20. Appl Environ Microbiol 66(6):2578–2588. doi:10.1128/AEM.66.6.2578-2588.2000

Chapter 4
Use and Overuse of Food Additives in Edible Products: Health Consequences for Consumers

Pasqualina Laganà, Emanuela Avventuroso, Giovanni Romano,
Maria Eufemia Gioffré, Paolo Patanè, Salvatore Parisi,
Umberto Moscato and Santi Delia

Abstract The aim of this chapter was to address the issue of uses and overuses of additives with consequent impacts on human health. Modern food additives have an important role in food production. At present, the number of individuals engaged in food primary production is dramatically decreasing; on the other hand, consumers seems to search for many food choices, including ready-to-eat foods, with annexed higher safety and hygiene standards (and food additives may be a concern). The most part of consumers have still confidence in the work of official commissions in defense of food safety; on the other hand, some people may find deem discrepancies between their own sense of security and the complex of laws or practices in defense of the human health. Main worries concern the enormous economic power of certain multinational groups in the field of food production; the absolute discretion of Committees to judge eligible or not eligible available researches on certain food additives; the choice of research methods and the lack of studies concerning synergist effects of different food additives.

Keywords Aspartame · Dye · European Food Safety Authority · Human health · Nitrate · Nitrite · Sulphite

Abbreviations

ADI	Admissible daily intake
EEC	European Economic Community
EUFIC	European Food Information Council
EFSA	European Food Safety Authority
FAO	Food and Agriculture Organization of the United Nations
FDA	Food and Drugs Administration
IARC	International Agency for Research on Cancer
JECFA	Joint FAO/WHO Expert Committee on Food Additives
SCF	Scientific Committee on Food
WHO	World Health Organization

© The Author(s) 2017
P. Laganà et al., *Chemistry and Hygiene of Food Additives*,
Chemistry of Foods, DOI 10.1007/978-3-319-57042-6_4

4.1   Use and Overuse of Food Additives and Impact on Human Health. An Introduction

Nowadays, the general matter of food additives can generate controversy and concerns among consumers because of the easy access to traditional and technologically advanced information media.

Actually, the use of food additives is not a modern discovery, linked to the technological progress in the field of industrial food productions. On the contrary, it is an ancient practice, aimed at making broader food resources. For example, salt addition or smoking treatments are systems lost in the mists of time (EUFIC 2006). A 1500 BC-Egyptian papyrus describes the practice of adding colouring and aromatic essences in order to make certain foods more desirable (Marmion 1991). The ancient Romans used saltpetre or potassium nitrate, spices and natural dyes to preserve and enhance the appearance of some foods. Baking soda is extensively used for centuries when speaking of baked products (Kitchen 2008); the same thing can be affirmed for thickening substances (preparation of sauces), or dyes such as cochineal for the preparation of sweets and liqueurs.

The European Council Directive No. 89/107/EEC gives a clear definition for a food additive: '*Any substance not normally consumed as a characteristic ingredient of food whether or not it has nutritive value, which are added to a food perform a technological purpose in the phases of production, processing, preparation, treatment, packaging, transport or storage, it may be reasonably expected to become, in it or its derivates, a component of such foods, either directly or indirectly*' (European Parliament and Council 1988).

Modern food additives today have an important role in food production. At present, the choice of available foods is wide, in qualitative and quantitative terms, and distributive services—supermarkets, shops, restaurants—are a notable part of the entire economic system in the Western world. We are witnessing a peculiar phenomenon: on the one side, the number of individuals engaged in food primary production is dramatically decreasing; on the other hand, consumers seems to search for many food choices, including ready-to-eat foods, with annexed higher safety and hygiene standards. All of these features should be affordable (EUFIC 2006). However, a growing number of people are lately willing to spend more to buy 'biologic' products because of the 'absolute absence' of 'dangerous' synthetic additives.

It has to be highlighted that food additives, including synthetic ones, perform a variety of functions which are now implicitly assumed by consumers. Therefore, additives are fundamental to preserve quality features and the palatability of foods, keeping them safe, nutritious, and appetising throughout the supply chain, from the farm to the table.

4.2 Food Additives and Food Safety

Food additives have to comply with strict mandatory requirements in the European Union and in other economic areas; in addition, their nature is related to edible foods without misleading intentions.

All additives substances must be tested and undergo tough safety assessment. In Europe, one of the main international assessment bodies is the Joint FAO/WHO Expert Committee on Food Additives (JECFA) by the Food and Agriculture Organization of the United Nations (FAO) and the World Health Organization (WHO). In addition, the Food and Drugs Administration (FDA) in the United States of America and Food Standards Agency of Australia and New Zealand are well considered when speaking of global food safety Authorities (Larsen and Richold 1999).

4.2.1 Food Additives and Food Safety. Evaluation Methods

All safety evaluations are made on available toxicological data, including investigations on animal and human models. This strategy allows the identification of the maximum level of additives without causing demonstrable toxic effects (also named: no-observed-adverse-effect-level, NOAEL). The next step is the definition of the 'admissible daily intake' (ADI) value for each substance.

The ADI is established taking into account the amount of the peculiar additive which can be taken with a normal daily diet, and it provides a wide margin of safety for all life stages. European regulations require that adequate researches have to be conducted to examine assumption levels of the population, and to correct any changes in consumer habits, in the case of excessive assumption. In this way, the occurrence of harmful effects is difficult even if ADI for a particular additive is exceeded, given a 100-factor wide margin of safety. In any case, if it should be systematic, the European Food Safety Authority (EFSA) may decide to decrease the levels of additives in food, or cut the allowed range of foods, if the maximum dosage may be systematically exceeded by certain groups of consumers (European Parliament and Council 2008). However, known consumer groups may not agree on such safety margins and publish articles and informative charts showing different evaluations.

4.2.2 The European Legislation

The European food market is not realizable without the harmonisation of National rules on food additives. In recent years, the European Economic Community (EEC) adopted the Framework Directive 89/107/EEC (1989) with the aim of defining evaluation criteria and making basis for peculiar documents such as Directives 94/35/EEC for sweeteners and 94/36/EC for colours (European Parliament and Council 1994a, b).

4.2.3   The Coding System

In Europe, each allowed food additive is indicated by an E followed by a number; historically, the approval was performed by the former Scientific Committee on Food (SCF); at present, SCF's responsibilities are transferred to the EFSA. The identification code, correlated with the International Numbering System (Sect. 2.2) is valid throughout the European Union territory and recognisable, in theory, on food labels by all consumers (EUFIC 2006).

4.2.4   Food Additives and Safety Concerns

Chemicals defined 'food additives' may have different origins (natural, synthetic, modified natural substances). For example, 'natural additives' include: vitamin C (E 300) and its derivatives (E 301-302-303), used as antioxidants; lycopene (E 160d); anthocyanins (E 163); vitamin B_2 (E 101); and curcumin (E 100), used for their colouring properties. Two common amino acids in the body, such as glutamic acid (E 620) and glycine (E 640) are used as flavour enhancers; citric acid (E 330) is an acidity corrector; pectin (E 440) is a thickener.

Some researchers have hypothesised since the 1970s that the increase in children who had behavioural and attention problems was caused by the modification in eating habits and quality of foods themselves. In particular, dyes provoked great interest and contradictory arguments (EUFIC 2006).

A recent review conducted in 2013 by the University of Belfast (Martyn et al. 2013) has reviewed all the studies concerning the relationship between consumption of additives and the onset of disorders of child behaviour in pre-school children. Researchers noted that, the 'additives/behavioral disorders' relation was demonstrated when researches were based on parental relationships, but the same thing did not happen without the evidence of similar data. Consequently, researchers conclude their discussion highlighting the need for a standardisation of diagnostic methods and a more holistic approach to research with relation to children' physiology and dyes at the same time (Martyn et al. 2013).

In addition, the common opinion of many European consumers about the possibility that additives may have adverse effects without evidence should be considered. Relationships between an immune system response and the ingestion of food additives have been demonstrated only in a few situations. The following list shows several of demonstrated allergenic effects:

- Dyes. Some sensitive subjects developed reactions to tartrazine (E 102) and carmine (E 120) with symptoms such as rash, nasal congestion and hives. The incidence is still very low (1–2 individuals affected every 10,000 people). The carmine has been rarely correlated with immunoglobulin E-mediated allergic reactions
- Sulphites. In detail, the following chemicals have been mentioned: inorganic sulphite additives (E 220–228), such as for example sodium sulfite, potassium bisulphite and metabisulphite containing sulfur dioxide (preservatives used to control microbial growth in fermented beverages and fruit juices, they can cause respiratory attacks in patients with asthma. Sulphites are also used in canned fish, frozen shellfish, nuts, in starch and potato flakes. In addition to respiratory symptoms, these chemicals can cause headache, cough and weakness, are typical ailments following heavy meals with shellfish and white wine. The presence of sulphites has been obligatorily mentioned on wine labels when the concentration is higher than 10 mg/kg of grapes because these chemicals are recognised as allergenic substances
- Monosodium glutamate. This substance is a flavour enhancer used in ready meals, sauces, soups, and Chinese specialties. It has been repeatedly reported to cause some side effects such as headaches and tingling, but scientific studies do not reveal this relationship
- Aspartame. It is a sweetener with almost no calories; for this reason, it is massively used in low-calorie foods. Its use has been questioned many times recently for multiple negative effects, none of which has been corroborated by official scientific data (EUFIC 2006)
- Nitrates and nitrites. Recently, the International Agency for Research on Cancer (IARC) has defined nitrates and nitrites as probably carcinogenic agents for humans. These two substances are not cancerous agents when taken individually; the problem is related with their transformation into N-nitrosamines, certified as carcinogenic agents after cooking (nitrates and nitrites should be contained in protein-rich foods), or after peculiar biochemical mechanisms. For example, sodium nitrite can turn into a general N-nitrosamine in acidic aqueous solution by means of the initial formation of nitrous acid (Eq. 4.1), the production of intermediate nitrosonium ion in equilibrium (Eq. 4.2), and the final reaction with a NHR_1R_2 amine (Eq. 4.3). The final product is the $R_1R_2N{-}NO$ nitrosamine, where R_1 and R_2 are different alkyl groups.

$$NaNO_2 \xrightarrow{H_2O^+} Na^+ + HNO_2 + H_2O \tag{4.1}$$

$$HNO_2 + H_3O^+ \leftrightarrow NO^+ + 2H_2O \tag{4.2}$$

$$NHR_1R_2 + NO^+ \xrightarrow[-H_2O]{} R_1R_2N - NO \tag{4.3}$$

Nitrates are found naturally in some vegetables such as beets, celery, turnips, spinach and in some drinking waters. They can be transformed into nitrite ions by the action of saliva during chewing. Subsequently, nitrites can combine with amines present in food proteins and form N-nitrosamines. For example, sodium nitrites are converted in the stomach (acid environment) into nitrous acid which easily combines with primary and secondary amines: the final product is an N-nitrosamine. This conversion can also be favoured at high temperatures (frying treatments).

Long-term use of nitrite in large doses has been associated with an increased cancer risk (target: stomach or oesophagus). As a result, consumers should be adequately advised with concern to foods that contain: potassium nitrite (E 249); sodium nitrite (E 250); and potassium nitrate (E 252). These foods are mainly represented by canned meat, sausages and industrial meat preparations.

Certain antioxidants such as vitamin C (E 300) or ascorbic acid and its derivatives such as potassium ascorbate (E 303) and sodium ascorbate (E 301) are able to inhibit the transformation into N-nitrosamines. Consequently, they are often used in combination with nitrates and nitrites (IARC Working Group 2010).

On the other side, nitrites are powerful preservatives: they are essential for their ability to prevent the growth of potentially fatal life forms such as pathogenic *Clostridium botulinum* (because of the production of a potent neurotoxin and related botulism syndrome). At present, legislations rule the use of nitrite in small quantities just for those foods where health risks associated to botulinum toxin are much greater than the chance of developing a tumor. The logical conclusion is to advise the consumer to a drastic reduction in the consumption of foods with added nitrates and nitrites, in favour of a larger consumption of fruits and vegetables (naturally rich in antioxidants and vitamins, natural inhibitors of N-nitrosamines).

4.2.5 Effectiveness of Safety Controls

Despite all assurances put in place by Health Authorities intended to reassure consumers about the harmlessness of permitted food additives, there is a creeping suspicion pervading a notable amount of the population. International bodies that oversee the use of additives constantly promote studies to check their safety.

Most of the work performed by experts' committees is spent on international studies submitted for review. Examined studies are conducted by both Producers food additives and independent laboratories (the FDA considers manufacturers' tests). Unfortunately, the citizens' trust in the European Scrutiny Committee on Food Security (SCF) work has been questioned in the past decade. One peculiar situation (Belpoggi 2013; Belpoggi et al. 2006; EFSA 2015; Millstone 2014; Soffritti 2007; Soffritti et al. 2005, 2007) has been observed when speaking of possible carcinogenic effects associated with aspartame, an artificial sweetener present in over 6000 products (paediatrics drugs are more of 500 different products).

4.2.6   *Conclusions*

Every consumer is in contact with food additives intentionally or unintentionally. All official Institutions in the world are responsible for the control of these substances with the aim of raising food safety and health, including the reliable evaluation of currently allowed additives for foods and related ADI values. Consumers have just to trust, and most of them have still confidence in the work of official Commissions, as the mechanisms adopted to safeguard our health seem sufficient. On the other hand, some people may find deem discrepancies between their own sense of security and the complex of laws or practices in defense of the human health.

What are the critical points that can undermine confidence in the Authorities and monitoring committees? First of all, citizens may consider with notable suspect the enormous economic power of certain multinational groups in the field of food production, with possible consequences on 'market laws' and consumers' health.

Another critical point concerns the absolute discretion of Committees such as the former SCF to judge eligible or not eligible available researches on certain food additives for peculiar purposes. Possible bias suspicion with concern to these choices can deeply undermine the confidence of taxpayers/consumers.

The third detectable critical point lays in the research methods (subjects: animals, humans, selected human groups such as children, pregnant women, individuals with acute or chronical diseases, etc.).

Finally, it should be noted that recent researches are mainly carried out on single food additives, with a little choice when speaking of synergist effects between two or more of these additives at the same time. For example, a study conducted by the Soil Association on E 133 (bright blue), E 104 (quinoline yellow), E 621 (monosodium glutamate) and E 951 (aspartame) seems to have demonstrated a neurotoxic effect. This effect is not apparently observed if mentioned additives are taken individually.

References

Belpoggi F (2013) Il commento dell'Istituto Ramazzini al verdetto di EFSA che ha mantenuto il livello di assunzione giornaliera di 40 mg/Kg di peso come limite di sicurezza. Istituto Ramazzini, Bologna. Available www.ramazzini.org/wp-content/uploads/2009/02/Aspartame-dicembre-2013-Belpoggi.pdf. Accessed 19 Dec 2016

Belpoggi F, Soffritti M, Padovani M, Degli Esposti D, Lauriola M, Minardi F (2006) Results of long-term carcinogenicity bioassay on Sprague-Dawley rats exposed to aspartame administered in feed. Ann N Y Acad Sci 1076:557–559. doi:10.1196/annals.1371.080

EFSA (2015) Oltre il 97% degli alimenti contiene residui di pesticidi nei limiti di legge. European Food Safety Authority, Parma. Available www.efsa.europea.eu/it/press/news/131210. Accessed 19 Dec 2016

EUFIC (2006) FONDAMENTI 06/2006. Additivi alimentari. European Food Information Council (EUFIC), Brussels. Available www.eufic.org/article/it/expid/basics-additivi-alimentari. Accessed 19 Dec 2016

European Parliament and Council (1988) Directive 87/107/EEC (1988) on the approximation of the laws of the Member States concerning food additives authorised for use in foodstuffs intended for human consumption. Off J Eur Comm L40:27–33

European Parliament and Council (1994a) European Parliament and Council Directive 94/35/EC of 30 June 1994 on sweeteners for use in foodstuffs. Off J Eur Comm L 237:3–12

European Parliament and Council (1994b) European Parliament and Council Directive 94/36/EC of 30 June 1994 on colours for use in foodstuffs. Off J Eur Comm L 237:13–29

European Parliament and Council (2008) Regulation (EC) No 1333/2008 of the European Parliament and of the Council of 16 December 2008 on food additives, lastly amended by Commis-sion Regulation No 2015/538 of 31 March 2015. Off J Eur Comm L 354:16–33 nessuno

IARC Working Group (2010) IARC Monographs on Evaluation of Carcinogenic Risks to Humans, Volume 94. World Health Organization, International Agency for Research on Cancer (IARC), Lyon

Kitchen L (2008) The baker. Murdoch Books, Sydney

Larsen JC, Richold M (1999) Report of workshop on the significance of excursions of intake above the ADI. Regul Toxicol Pharmacol 30(2 Pt 2):S2–S12

Marmion DM (1991) Handbook of U.S. colorants: foods, drugs, cosmetics, and medical devices, 3rd edn. Wiley, New York

Martyn DM, McNulty BA, Nugent AP, Gibney MJ (2013) Food additives and preschool children. Proc Nutr Soc 72:106–109. doi:10.1017/S0029665112002935

Millstone E (2014) EFSA on aspartame, January and December 2013. University of Sussex, Brighton. Available www.laleva.org/it/docs/Millstone_EFSA_Aspartame_9jan2014. Accessed 19 Dec 2016

Soffritti M (2007) Aspartame: Soffritti responds. Environ Health Perspect 115(1):A17

Soffritti M, Belpoggi F, Degli Esposti D, Lambertini L (2005) Aspartame induces lymphomas and leukaemias in rats. Eur J Oncol 10:107–116

Soffritti M, Belpoggi F, Tibaldi E, Esposti DD, Lauriola M (2007) Life-span exposure to low doses of aspartame beginning during prenatal life increases cancer effect in rats. Environ Health Perspect 115(9):1293–1297. doi:10.1289/ehp.10271